FISH CYTOGENETIC TECHNIQUES

Ray-Fin Fishes and Chondrichthyans

FISH CYTOGENETIC TECHNIQUES

Ray-Fin Fishes and Chondrichthyans

Editors

Catherine Ozouf-Costaz
CNRS, UMR 7138 – Evolution Paris-Seine
IBPS, UPMC
Paris
France

Eva Pisano
DISTAV – Dipartimento di Scienze della Terra, Ambiente, Vita
University of Genova
Genova
Italy

Fausto Foresti
UNESP – São Paulo State University
Institute of Biosciences
Department of Morphology
Botucatu, SP
Brazil

Lurdes Foresti de Almeida Toledo
Departamento de Genética e Biologia Evolutiva
Instituto de Biociências
Universidade de São Paulo
São Paulo, SP
Brazil

CRC Press
Taylor & Francis Group
Boca Raton London New York

CRC Press is an imprint of the
Taylor & Francis Group, an **informa** business
A SCIENCE PUBLISHERS BOOK

Cover Illustrations

From top to bottom and from left to right:

Meiotic chromosomes (pachytene) in *Eigenmannia* sp. where central element proteins of the synaptonemal complex (SCYP1) were detected (photograph C. Araya Jaime); meiotic chromosomes (diplotene) in *Characidiumgomesi* where the lateral element proteins of the synaptonemal complex (SCYP3) were detected (photograph C. Araya Jaime); mitotic chromosomes and interphase nucleus of *Trematomusbernacchii* showing localization of Immunoglobulin heavy chain genes by FISH (photograph: C. Ozouf-Costaz and E. Pisano); mitotic chromosomes of *Trematomuspennellii* after FISH of one family of DIRs transposable elements (photograph J. Auvinet and C. Ozouf-Costaz).

CRC Press
Taylor & Francis Group
6000 Broken Sound Parkway NW, Suite 300
Boca Raton, FL 33487-2742

First issued in paperback 2019

ISBN-13: 978-1-4822-1198-6 (hbk)
ISBN-13: 978-0-367-37755-7 (pbk)

Visit the Taylor & Francis Web site at
http://www.taylorandfrancis.com

and the CRC Press Web site at
http://www.crcpress.com

Foreword

Biology of the XXth century went deeper and deeper into the characterization and use of biological structures for various scientific purposes. Chromosomes have been observed by Flemming in 1880 but cytogenetics started flourishing only by the late fifties. However in research institutions, zoology and histology were slowly but surely replaced by cellular biology, soon replaced by molecular biology. During the sixties, cytogenetics could wrap itself into modernity, soon apparently supplanted by molecular genetics that, since then, has been considered as the only show in town for long. Some stormy debates continue to be held today, carrying the fear that molecular approaches to the diversity of life would erase more integrative approaches like comparative anatomy. Scientists often behave as if one level of observation should necessarily destroy and replace the preceding larger one: the smaller the scale of observation is, the better. This is scientifically unsound. Whatever the kind of question we ask during our biological investigations, Science still needs all levels of observation—populational, behavioural, anatomical, histological, cellular, chromosomal, genetic and molecular—to make sense of organismic diversity, functioning and history. Indeed, all these levels are not mobilized the same way according to the type of investigation. But research potential at each level must be well alive, involving a vast variety of disciplines and questions if we plan to practice good science. The most fruitful collaborations we experience in our professional lives is a matter of dialog between these levels. Cytogenetics fully plays this game. It is sociology and economy of scientific communities that explain why some levels of biological integration become obsolete, not science itself. We are still hardly able to put the same names on muscles of a catfish on one hand, and a perch on the other. There are still debates about what a parietal bone is. Similarly, one could think that we know everything about chromosomes, just because complete genome DNA sequences are just so easy to obtain today. Our race to the smaller is not a matter of scientific soundness. It is a matter of technophilia embedded into a society of mass consumption. Technological novelties are spectacular but make us forget our scientific goals and questions. There is no privileged level of integration for all scientific questions. The real factor that characterizes the kind of science we're practicing is more a matter of kind of reasoning. All evolutionary sciences (and, largely, all of biology) function according to two distinct, parallel reasonings:

- Structural sciences logically organize what exists into nested sets and name them. They answer questions of the type: "what is that?" and "where does it come from?" These sciences (comparative anatomy, comparative cytogenetics, descriptive embryology, paleontology, systematics, molecular phylogenetics, etc.) are historical. They must rationally explain structures observed in a time period that is not the one of the organism itself (except perhaps in the case of descriptive embryology), but the time of the species and phylum histories. The cause-consequence relationship is not embedded into the organismic time, it is embedded into history. For example, reconstruction of an ancestral karyotype belongs to this first type of procedure.
- Sciences of processes exhibit cause and effect relationships in the time frame of biological processes and answer the question "how does it work?" In biology (molecular genetics, embryology, physiology, population genetics, ecology, etc.), using experimental proof, those sciences aim to explain the underlying mechanisms of the biological phenomena, and the phenomenon of biological evolution itself, either in the time of the organism (population genetics, cytogenomics, causal embryology, molecular genetics), or in the course of species' history (ecology, population genetics). Studying the role of chromosomes in genetic expression falls under this second instance.

Cytogenetics played the right game through this epistemological dichotomy. Explaining how the genome is functioning is useful for structure comparisons in order to reconstruct the history of genomes, or species. Cytogenetics has also been successful in adapting itself to the rise of DNA technology, and more generally, all molecular younger technologies—the present book will probably convince the reader. Cytogenetics has been able to incorporate them and evolve to reveal a richer scope of phenomena. It is clear through this book that cytogenetics is today technically flexible and potentially fruitful to historical sciences (revealing biological patterns) and etiological sciences (enlighting ongoing biological processes). Cytogenetics still has a long scientific life in front.

Guillaume Lecointre
ISYEB, UMR 7205, Muséum National d'Histoire Naturelle
Département Systématique et Evolution
CP39, Muséum National d'Histoire Naturelle
57 rue Cuvier 75231 PARIS cedex 05, France

Preface

This book is written for scientists and students who need detailed protocols that allow for the preparation of chromosomes from ray-finned fishes (actinopterygians) and cartilaginous fishes (chondrichthyans), not only for karyotype establishment but also for elucidating chromosome structure and gene content. In this book, specialists from eight countries make available their expertise in fish cytogenetics by providing their best protocols for chromosome preparation and the main molecular cytogenetic techniques that they currently use. The numerous applications address both general questions (comparative cytogenetics, evolutionary studies at different levels of genome organization) and more specific questions (reproductive biology, agronomy, veterinary medicine, gene mapping, assembling genome sequences, etc.).

Compared to other vertebrates, the chromosomes and karyotypes from actinopterygians and chondrichthyans are often difficult to obtain. Except for domesticated species, this entails the careful collection of the specimens in their usual environment, and their live maintenance on board ships or in laboratories in conditions suitable to their physiology, until tissue sampling for chromosome preparations can be undertaken. The high species diversity within each of these two fish groups results in a wide range of techniques used for cytogenetic purposes, so that, unlike with birds and mammals, it is impossible to describe a standard protocol for chromosomal preparation. Each new fish group being studied requires the development of specific protocols, which must be adjusted to field or laboratory conditions. Therefore a major part of this book is dedicated to diverse methods for *in vivo* or *in vitro* chromosome preparations. Chromosomes of the highest quality are a prerequisite for the success of further techniques, such as chromosome "banding" and fluorescence *in situ* hybridization (FISH), as currently used in classical and molecular cytogenetics of fishes.

For each of the 20 chapters, some unpublished tips and tricks are provided, as well as keys for the success of the described methods. When alternative methods are provided in different chapters, the reader is encouraged to make appropriate choices according to the goals, and in consideration of the available material and equipment. In chapters dealing with molecular cytogenetics, the reader should also be aware that methods and protocols in this field (such as FISH, CGH, GISH, chromosome painting, fiber-FISH) have advanced by progressively adapting the techniques previously set up for mammals. Some major technical difficulties are often encountered, due to the generally small size of the fish chromosomes, and because of their fragility during the denaturation process. For example, while FISH has been used on fish chromosomes for two decades, chromosome painting could only be performed with a reasonable rate of success on fishes until very recently.

All chapters follow the same outline: (1) a description of the general principles of the method, indicating on which fish species/group the method has proved to be efficient; (2) a brief enumeration of the applications; (3) a step-by-step description of the protocol; (4) a "troubleshooting" section, containing supplementary information and recommendations for modifying the protocol in the case of failure; (5) the list of material, equipment, chemical, reagents, stock and working solutions; (6) the cited references.

Annexes 1, 2 and 3 provide a general glossary, a general list of abbreviations, and the protocols for preparing the most commonly used stock and working solutions in cytogenetics.

The editors would like to warmly thank all the contributors for their collaboration and patience during the long period necessary to finish the book, and O. Coriton and P.A. Hulley for kindly reviewing several chapters. They also wish to specially honour the memory of Dr. Maria del Carmen Mühlman (co-author of Chapter 17), who undertook pioneering work on fish chromosome micro-dissection and chromosome painting.

Catherine Ozouf-Costaz
Eva Pisano
Fausto Foresti
Lurdes Foresti de Almeida Toledo

Contents

1

Teleost Fish Handling and Transport under Reduced Stress Conditions

B. Aupérin[1] and *J.-F. Baroiller*[2,*]

[1]INRA, UR1037-LPGP, Fish Physiology and Genomics, Campus de Beaulieu, F35000 Rennes, France.
e-mail: benoit.auperin@rennes.inra.fr

[2]CIRAD, UMR116 ISEM, Campus International de Baillarguet, TA B-116/A, 34398 Montpellier cedex 5, France.
e-mail: baroiller@cirad.fr

*Corresponding author

Principles

With over 32,000 species, teleost fish have adapted to a tremendous range of aquatic habitats. At least 200 of them are used for aquaculture all over the world and even more are handled and transported for the ornamental fish market, as well as for bait or research purposes, including cytogenetic studies.

Scientific information on stress induced by handling and transport of the fish is very scarce, especially concerning wild species. Technical journals refer to random investigations carried out only once, and therefore these results can only give some initial ideas of the problem. Anyway, attention to the quality of the packaging and transport certainly decreases mortality. Schütz (2003) showed that transport mortality was significantly higher for non-CITES (Convention on International Trade in Endangered Species) species than for CITES ones for mammals, birds, reptiles and amphibians possibly because of greater care in handling CITES animals during transport, since public authorities pay closer attention to these species.

Fish transport can be divided into four steps: 1) capture, 2) loading, 3) shipment management and 4) recovery. Sensitivity of fish to these different steps differs greatly according to the species: for example, initial capture and loading appear to be very stressful for salmonids rather than the transport itself (Barton et al. 1980; Specker and Schreck 1980). Transport *per se* is stressful for carp (Svobodova et al. 1999).

Applications and Limitations

Before proceeding to any fish shipment from one site to another, especially for wild species, it is necessary to be informed of the latest specific modalities for access and transfer of fish species (licences, authorizations, veterinary certificates, quarantines…) according to the specific conservation, sanitary and transportation rules of the two countries (country of origin and final destination).

Fish's needs for water quality are dependent upon the species: Carps, tilapias or catfish can survive levels of oxygen deficit and suspended solids that would be lethal for salmonids.

Although it is difficult to generalize optimal conditions, we will give some basic principles here as a guideline.

Protocols

1. Capture or collection of the fish

This very stressful step can be associated with wounds, change of water quality, poor water quality (low oxygen, high temperature) and overcrowding. McDonald et al. (1993) recommended various ways by which initial disturbance can be minimised. There is very limited experimental data on the validity of most of these procedures, but all of them should be tested for each fish species. Special care should be given to fish collection because external damage will weaken or even lead to the death of the fish. When a fish has been captured in a net, it has to be directly transferred in a container with water from the collection site, by keeping the net in the water, in order to transfer the fish without it emerging from the water. Injured specimens have to be released. The best method will be to stock the fish for several days before shipping them and regularly changing the water (using the water from the collection site when possible).

Starvation before shipping

Adult fish should be starved before transportation to allow the gut to clear, thus decreasing the bacterial and faecal load placed on any holding and transport system. The length of starvation depends upon the species, size of the gut, and of the water temperature; 24 hours will be the minimum period (increase this period with low temperatures). Because of their very limited reserves, larval fish should not be starved or if so, only for a short period (maximum 24 hours).

N.B.: Direct anaesthesia in the tank of storage has also been recommended for hybrid striped bass.

Conditioning the fish before shipping

Schreck et al. (1995) have shown that salmon handled before transport appeared to recover more rapidly than fish which had not been conditioned.

Minimising as far as possible the duration of any stress before shipping

The duration of the fish cortisol response is positively correlated with the duration of the stress itself (Pickering and Pottinger 1989).

Use of heat shock protein exposure prior to transport

Exposure to heat shock protein in the water before transportation significantly reduces mortality in trial transportations of salmon smolts and sea bass (B. Aupérin, unpublished data).

2. Loading

For many species, loading the fish into the container is the most stressful step of the transport process (Miles et al. 1974; Specker and Schreck 1980). It is associated with alterations of both behaviour and physiological state (Mazeaud et al. 1977), which may lead to mortality (Wedemeyer 1976). In chinook salmon, plasma cortisol levels increase after loading and then decrease during the period of the transport (Maule et al. 1988).

A critical factor in loading stress is the trauma associated with the method of transfer (Jeon et al. 2000a; Jeon et al. 2000b). Netting out and lifting within the net to place the fish in the container is the most widely used method of transfer and is often associated with both skin trauma and cortisol levels increase (Helfrich et al. 2001).

Fish skin is an extremely fragile barrier and any trauma on its surface, even without removing scales, will cause significant osmotic disturbance and will be a way for pathogen invasion. Thus, wherever possible, fish should be transferred within the water medium or with soft, uncoated, knotless nets (Carmichael et al. 2001).

3. Transport management

Special care should be paid to:

The water quality deterioration

Excretory products such as ammonia and carbon dioxide are dangerous for fish. In water, the equilibrium between the ionised ($NH4^+$ and $NH3^+$) and unionised forms of ammonia depends upon water pH and temperature. Increase of water pH and temperature increases non-ionised ammonia concentration (Emerson et al. 1975) which is highly toxic for fish. High carbon dioxide concentration decreases water pH and the toxicity of ammonia.

The water temperature

When possible it is recommended to reduce the water temperature and to transport fish in an insulated box even if the overall temperature can increase during transport.

The rate of chilling and the lethal temperature must be determined for each species. For carp, cooling fish with water from 25°C to 7°C in 4 hours is claimed to increase the survival, whereas dropping fish directly into such cold water can in itself cause losses and does not reduce subsequent transport stress.

Cooling the hauling tank water reduces fish metabolism (ammonia accumulation decreases) and activity. For red porgy a temperature drop from 25°C to 20°C reduces metabolism by about 50% and increases the carrying capacity fourfold (Takeda et al. 1989). For salmonids, cooling hauling tank water by 1°C will allow an increase in the fish loading density of about 10% (Wedemeyer 1992).

The oxygen availability

For a given fish species, water temperature and fish weight are the most important factors related to oxygen demand during transport. The lower the temperature is, the lower the oxygen consumption and the higher the dissolved oxygen are. Again it is difficult to provide values. However, for example, for silver catfish (2 g fingerlings), an oxygen deficit was observed at 20°C and 25°C at the load density of 168 g/l after 24 hours of transportation (Golombieski et al. 2003). Cod, for example, can be transported at density as high as 540 kg/m^3 water for 48 h (Staurnes et al. 1994). For a species and a given total weight in the hauling bag, small fish use more oxygen than large fish.

The carbon dioxide accumulation in water

When fish are transported in a sealed container (plastic bag, for example), carbon dioxide increases in the water and can reach toxic levels: 15 mg/L is the highest limit for rainbow trout (see Berka 1986 for other values). High level of carbon dioxide in water decreases its diffusion from the blood to the water, which in turn decreases the ability of haemoglobin to transport oxygen (the Bohr effect), decreases the maximum oxygen binding capacity of blood (the Root effect) and increases blood acidity (hypercapnia). This toxic effect of CO_2 on fish is difficult to predict and the toxic concentration value decreases when the oxygen concentration in the water decreases (Westers 2001). Moreover, increase of CO_2 concentration induces a decrease of the water pH which decreases the toxicity of ammonia accumulated during the transport.

The space allowance

Wedemeyer (1992) wrote: "the maximum weight of fish that can be routinely transported in a particular fish distribution unit must usually be determined by experience because it depends on variables such as the physiological condition of the fish to be hauled, the efficiency of the aeration system, and the chemical composition of the water". So, in defining the best transport conditions for a fish we must take into account not only the species, but also the water quality before loading, the age, the size and the conditions of the specimen. In addition, we have to consider the length of the journey, the temperature and the method of transport. For example, aggressive species such as Siamese fighting fish and large male tilapias have to be transported individually. Instead, fish like zebrafish and salmon smolts can be safely transported in batches of many thousands as long as water quality is correct.

To decrease activity, stress level and level of consciousness of the fish, it is possible to use anaesthetics at very low concentration. Several anaesthetics are available (methane tricaine sulphonate, benzocaine, metomidate, phenyl ethyl alcohol, clove oil) and are used to sedate fish during handling. MS222 has been widely tested in the USA as a transport facilitating anaesthetic and used with success to reduce transport stress and mortality of bluegills (Webb 1958), threadfin shad (Collins and Hulsey 1963) and gizzard shad (Anderson 1968). In EU no anaesthetic has drug licensing agreement at least for human consumption, but several are used in experimental facilities (clove oil, phenyl ethyl alcohol).

The right dose of the anaesthetic should be tested for each species before actually transporting because of the variability in specific sensitivity.

As a general rule and as far as possible, fish should be loaded in the container several hours before the transportation in well oxygenated water and left as undisturbed as possible to recover from the capture and loading stress. Just before container sealing, water should be changed (same water quality as before) several times without drying the fish. After that and before sealing the container, oxygen should be introduced into the bag (1/3 water, 2/3 oxygen). Water used should be as clean as possible: low ammoniac concentration and low quantity of suspended material. However, some species possess a dual type of respiration, including air breathing (Bettas, Gouramis, *Colisas, Corydoras, Osteoglossidae*...). For these species pure oxygen has to be replaced by air.

For bag transport

It is always reasonable to put the first bag into a second bag to avoid any leak of water and oxygen. The corners of the two bags can be crimped, in order to avoid the fish getting caught in the corners during shipment.

Some fish species have dorsal and/or pectoral fins with very sharp first rays (*Corydoras*, some Cichlids and catfish species...) and therefore in order to avoid bag perforation, the rays can be either wrapped with plastic tubes or cut under anaesthesia.

4. Recovery

Recovery after transport before experiments, can take several days (Bandeen and Leatherland 1997; Carmichael et al. 1984) and certainly increases subsequent survival (Jonssonn et al. 1999). Measurement of plasma cortisol levels of chinook salmon recovering from the stress of transportation suggested complete recovery after 24 hours. However other physiological data and results of challenge tests suggested that additional time may be necessary for full recovery (Maule et al. 1988). Studies with juvenile striped bass and walleye fry have shown survival to be inversely related to transport time (Pitman and Gutreuter 1993). For largemouth bass, Carmichael et al. (1984) demonstrated a correlation between the length of the haul and the severity of plasma chloride decrease during the four days after the haul was completed. In the same study the authors showed that mortalities occurring within four days of the transport can be reduced by pre-transport treatment (72 h fasting and 10 days prophylactic disease treatment).

In addition to the direct effects of transport, an animal's immune response can be compromised and lead to health problems during quarantine.

Receiving a shipment of fish

After transportation, it is usually necessary to change the water containing high concentrations of toxic material: ammoniac, suspended solids, CO_2 with new water having a very similar quality (pH, ions composition, hardness), well oxygenated and with low suspended solids. For this, the receiving tank or aquarium has to be prepared at least a few days before the arrival of the fish, in order to get a stable water quality, and especially stable pH and temperature. Check the water quality regularly.

In order to avoid thermal shock, unopened bags containing the fish will be let floating in the receiving water to allow a slow acclimatization. To avoid further stress, avoid handling the fish as much as possible. when opening the bag. If some fish die, remove them as quickly as possible. Feeding should gradually be started again (a 12 h starvation period is recommended).

Wild fish especially have to be left in a quiet place, preferably not in an aquarium where the fish can be disturbed and exposed to a bright light. Such light is usually not necessary because it will not meet the former living conditions of the fish (water is not always clear). The use of an opaque box is probably better. It is necessary to let the fish recover for several days.

Sampling for cytogenetic analysis

Direct chromosome preparations. Fish can be used for removing head kidney, spleen or gonads for direct mitotic and meiotic direct chromosome preparations: in this case, they must receive a lethal anaesthetic treatment.

In vitro chromosome preparations. Fish can also be used for lymphocyte culture. In this case, the effects of stress must be reduced as much as possible, because cortisol may inhibit the cell growth in the culture. To reduce the stress of handling for cytogenetic (and other biological) sampling, care should be taken with the fish living conditions after transportation. Furthermore, to reduce the stressing effects of netting, *in vivo* treatments and manipulations of fish out of the water, an anaesthetic is generally used. Fish can be netted, transferred in a small tank containing water with the anaesthetic. Tests should be made to adjust the anaesthetic concentration for each given species, because fish response to the anaesthetic is variable according to type of anaesthetic, species, temperature and the size of specimens. 3 ml/10 L of phenoxy-ethanol is enough for rainbow trout regardless of the fish size when the temperature is below 16°C. When possible (when tanks are not too big) fish can be anaesthetized directly in the tank where they are reared (Auperin et al. 1997; 1998). Usually, the same anaesthetic concentration as before or even lighter concentrations can be used, especially for treatments *via* injection.

Fin clips. They can also be done for fibroblast culture after anaesthetizing fish.

Blood sampling for lymphocyte culture. Because stress can cause a decrease of the mitotic activity, when blood is used for lymphocyte culture, precautions must be taken. Blood samples can be obtained without anaesthetic when a small number of fish are to be sampled, or when fish should be kept alive. After removal of all the fish from water, all fish are sampled in a very short time (less than 1 min). To be done well, this technique requires one operator per fish (see example on tilapia in Auperin et al. 1997). After the first netting in one tank, one should wait at least one day before using other fish from the same tank in order to be sure that the fish are not stressed further.

If it is not possible to have one operator per fish, a higher concentration of anaesthetic can be used. This will reduce the stress (especially the increase in cortisol concentration and, probably with less success, in epinephrine concentration). Anaesthetic concentration should be adjusted to obtain the sedation in less than 30s (especially for tropical species) following the opening of the rearing tank. After this sedation, fish can be sampled for several minutes. Since these high anaesthetic concentrations may kill the fish, care should be taken to the duration during which the blood can be successfully sampled after death. Again, an empirical test should be done to know the concentration of anaesthetic to use and the length of time during which the fish can be sampled (Auperin et al. 1997). Fish slaughtering with a blow on the head can also solve the single operator problem, but as above, the sampling duration after animal death has to be assessed.

Troubleshooting

Improving the transportation process

The occurrence of shipments with relatively high mortality suggests an inappropriate temperature, overstocking, poor condition of animals before shipment, poor packaging, inadequate supply of oxygen or transport of too young fish without feeding.

To improve the transportation quality, some basic principles are:

1) Reduce fouling and faecal material by withdrawn 24 to 72 hours, depending on water temperatures, prior to transport in closed systems.
2) Reduce the stress (to reduce metabolism, accumulation of ammonia and carbon dioxide and oxygen requirements).

    - Avoid unnecessary manipulation. The ability of wild animals to withstand the rigours of transportation is highly affected by the pre-transportation handling.
    - Reduce disturbance among fish. Fish from different groups should not be mixed in common containers during transport because severe stress can be induced by the hierarchies and fighting.
    - Use low doses of anaesthetics prior to and during transportation.
    - Reduce the water temperature (Barton and Peter 1982; Carmichael et al. 1984; Robertson et al. 1988; Wedemeyer 1972) and avoid abrupt water temperature change during transportation. Higher water temperature reduces the oxygen solubility with subsequent oxygen demand by fish. Ice or gel packs can be used during shipment to decrease or/and maintain a lower temperature. Conversely, for tropical species or for shipment during Fall, Winter or early Spring, heat packs can be taped to the inside bottom of the styrofoam box below a few layers of newspaper. After activation (exposure to oxygen), the heat pack will provide 20–40 h of heat.
    - Manipulate the water osmolarity. Attempts have been made to reduce the effects of stress on ion balance and ameliorate the loss of

homeostasis, by the use of diluted salt solutions (0.5–1.0% NaCl and in soft water $CaCl_2$ to bring hardness up to 50 mg/l) during transport of fish (Hattingh et al. 1975). The beneficial effects of such treatment have been demonstrated for several species (Hattingh et al. 1975; Long et al. 1977). This treatment has the effect of lowering the osmotic gradient between plasma and the environment, thus reducing the energy cost for osmoregulation (Redding and Schreck 1983).
- Buffer the water to confer benefits in situations where alkalinity is low and there is no ventilation in the transport containers.

Great variability exists in the efficiency of these different treatments in various species, thus resulting in a lack of consensus on an optimal standard transport method (Barton and Peter 1982).

References

Anderson, R.O. 1968. Transport of grizzard shad. Progressive Fish-Culturist 30: 184.

Auperin, B., J.F. Baroiller, M.J. Ricordel, A. Fostier and P. Prunet. 1997. Effect of confinement stress on circulating levels of growth hormone and two prolactins in freshwater-adapted tilapia (*Oreochromis niloticus*). Gen. Comp. Endocrinol. 108: 35–44.

Auperin, B., L. Goardon, A. Quemeneur, J.L. Thomas, J. Aubin, C. Valotaire, Y. Rouger and G. Maisse. 1998. Etude préliminaire sur l'utilisation de l'Aqui'S comme anesthésique pour la manipulation et l'échantillonnage de truites arc-en-ciel (*Oncorhynchus mykiss*) et de truites fario (*Salmo trutta*). B Fr. Pêche Piscic 350-351: 291–301.

Bandeen, J. and J.F. Leatherland. 1997. Transportation and handling stress of white suckers raised in cages. Aquacult. Int. 5: 385–396.

Barton, B.A. and R.E. Peter. 1982. Plasma cortisol stress response in fingerling rainbow trout, *Salmo gairdneri* Richardson, to various transport conditions, anaesthesia, and cold shock. J. Fish Biol. 20: 39–51.

Barton, B.A., R.E. Peter and C.R. Paulencu. 1980. Plasma cortisol levels in fingerling rainbow trout (*Salmo gairdneri*) at rest, and subjected to handling, confinement, transport and stocking. Can. J. Fish. Aquat. Sci. 37: 805–811.

Berka, R. 1986. Le transport des poissons vivants. Etude de synthèse. *In*: EIFAC 17 TECH. PAP no. 48, Rome (Italy).

Carmichael, G.J., J.R. Tomasso, B.A. Simco and K.B. Davis. 1984. Characterization and alleviation of stress associated with hauling largemouth bass. T. Am. Fish. Soc. 113: 778–785.

Carmichael, G.J., J.R. Tomasso and T.E. Schwedler. 2001. Fish transportation. pp. 641–660. *In*: Wedemeyer, G.A. (ed.). Fish Hatchery Management, second edition. American Fisheries Society Bethesda, Maryland.

Collins, J.L. and A.H. Hulsey. 1963. Hauling mortality of threadfin shad reduced with MS-222 and salt. Prog. Fish. Cult. 25: 105–106.

Emerson, K., R.C. Russo, R.E. Lund and R.V. Thurston. 1975. Equilibrium calculations: effects of pH and temperature. J. Fish. Res. Board Can. 32: 2379–2383.

Golombieski, J.I., L.V.F. Silva, B. Baldisserotto and J.H.S. da Silva. 2003. Transport of silver catfish (*Rhamdia quelen*) fingerlings at different times, load densities, and temperatures. Aquaculture 216: 95–102.

Hattingh, J., F. Fourie and J. Van Vuren. 1975. The transport of freshwater fish. J. Fish Biol. 7: 447–449.

Helfrich, L.A., C. Liston and D. Weigmann. 2001. Evaluation of a helical pump to transport chinook salmon and striped bass. Aquaculture 2001: 288.

Jeon, J.K., P.K. Kim, Y.J. Park, J.G. Myoung and J.M. Kim. 2000a. Changes of serum cortisol concentration and stress responses in coho salmon (*Oncorhynchus kisutch*) to netting. J. Korean Fish. Soc., Pusan. 33: 115–118.

Jeon, J.K., P.K. Kim, Y.J. Park, J.G. Myoung and J.M. Kim. 2000b. Stress responses of coho salmon, *Oncorhynchus kisutch*, to transport in fresh water or salt water. J. Korean Fish. Soc. Pusan. 33: 119–123.
Jonssonn, S., E. Braennaes and H. Lundqvist. 1999. Stocking of brown trout, *Salmo trutta* L.: effects of acclimatization. Fisheries Manag. Ecol. 6: 459–473.
Long, C.W., J.R. McComas and B.H. Monk. 1977. Use of salt (NaCl) water to reduce mortality of chinook smolts, *Oncorhynchus tshawytscha*, during handling and hauling. Mar. Fish. Rev. 39: 6–9.
Maule, A.G., C.B. Schreck, S. Bradford and B.A. Barton. 1988. Physiological effects of collecting and transporting emigrating juvenile chinook salmon past dams on the Columbia river. T. Am. Fish. Soc. 117: 245–261.
Mazeaud, M.M., F. Mazeaud and E.M. Donaldson. 1977. Primary and secondary effects of stress in fish: some new data with a general review. T. Am. Fish. Soc. 106: 201–212.
McDonald, D.G., M.D. Goldstein and C. Mitton. 1993. Responses of hatchery-reared brook trout, lake trout, and splake to transport stress. T. Am. Fish. Soc. 122: 1127–1138.
Miles, H.M., S. Loehner, D.T. Michaud and S.L. Salivar. 1974. Physiological responses of hatchery reared muskellunge (*Esox masquinongy*) to handling. Trans. Am. Fish. Soc. 103: 336–342.
Pickering, A.D. and T.G. Pottinger. 1989. Stress responses and disease resistance in salmonid fish: Effects of chronic elevation of plasma cortisol. Fish Physiol. Biochem. 7: 253–258.
Pitman, V.M. and S. Gutreuter. 1993. Initial poststocking survival of hatchery-reared fishes. N. Am. J. Fish. Manage. 13(1): 151–159.
Redding, J.M. and C.B. Schreck. 1983. Influence of ambient salinity on osmoregulation and cortisol concentration in yearling coho salmon during stress. T. Am. Fish. Soc. 112: 800–807.
Robertson, L., P. Thomas and C.R. Arnold. 1988. Plasma cortisol and secondary stress responses of cultured Red Drum (*Sciaenops ocellatus*) to several transportation procedures. Aquaculture 68: 115–130.
Schreck, C.B., L. Jonsson, G. Feist and P. Reno. 1995. Conditioning improves performance of juvenile Chinook salmon, *Oncorhynchus tshawytscha*, to transportation stress. Aquaculture 135: 99–110.
Schütz, C. 2003. Transport Losses of CITES-protected and Non Protected Animal Species: Report on Research and Development Project 89986340 of the Federal Agency for Nature Conservation "Identification of Transport Sensitive CITES-protected Species", BfN-Skripten-Deutschland Bundesamt für Naturschutz, 104 pp.
Specker, J.L. and C.B. Schreck. 1980. Stress responses to transportation and fitness for marine survival in coho salmon (*Oncorhynchus kisutch*) smolts. Can. J. Fish Aquat. Sc. 37: 765–769.
Staurnes, M., T. Sigholt, H.P. Pedersen and T. Rustad. 1994. Physiological effects of simulated high-density transport of atlantic cod (*Gadus morhua*). Aquaculture 119: 381–391.
Svobodova, Z., P. Kalab, L. Dusek, B. Vykusova, J. Kolarova and D. Janouskova. 1999. The effect of handling and transport on the concentration of glucose and cortisol in blood plasma of common carp. Acta Veterinaria Brno. 68: 265–274.
Takeda, T., T. Tsuji and Y. Itazawa. 1989. Facilitation of housing capacity for red porgy by drop in water temperature. Nippon Suisan Gakk. 55: 1011–1015.
Tomasso, J.R., K.B. Davis and N.C. Parker. 1980. Plasma corticosteroid and electrolyte dynamics of hybrid striped bass (white bass x striped bass) during netting and hauling. Proceedings of the World Mariculture Society 11: 303–310.
Webb, R.T. 1958. Distribution of bluegill treated with tricaine methnesulfonate (MS-222). Progressive Fish-Culturist. 20: 69–72.
Wedemeyer, G.A. 1972. Some physiological consequences of handling stress in the juvenile coho salmon (*Oncorhynchus kisutch*) and steelhead trout (*Salmo gairdneri*). J. Fish. Res. Bd. Can. 29: 1780–1783.
Wedemeyer, G.A. 1976. Physiological response of juvenile coho salmon (*Oncorhynchus kisutch*) and rainbow trout (*Salmo gairdneri*) to handling and crowding stress in intensive fish culture. J. Fish. Res. Bd. Can. 33: 2699–2702.
Wedemeyer, G.A. 1992. Transporting and handling smolts. World Aquac. 23: 47–50.
Westers, H. 2001. Production. pp. 31–89. *In*: Wedemeyer, G.A. (ed.). Fish Hatchery Management. American Fisheries Society, Bethesda, Maryland.

2

Storage of Karyotyped Voucher Specimens and their Molecular Identification

A. Dettai[1,*] and *P. Pruvost*[2]

[1]Institut de Systématique Evolution Biodiversité, UMR7205 ISYEB MNHN-CNRS-UPMC-EPHE, Département Systématique et Evolution, Muséum national d'Histoire naturelle, 75231 Paris cedex 05, France.
e-mail: agnes.dettai@mnhn.fr

[2]UMR720 BOREA MNHN-CNRS-UPMC-IRD, Département Milieux et Peuplements Aquatiques, Muséum national d'Histoire naturelle, 75231 Paris cedex 05, France.
e-mail: patrice.pruvost@mnhn.fr

*Corresponding author

Principles

Fish are currently the only vertebrate group where several hundred new species are still described each year (Fig. 1). One third of the currently recognized species have been described in the last 12 years (Fishbase 2002; 2014). The reassesment using modern approaches of species that were thought to be well known resulted in the discovery of many new species (Thomas et al. 2014; Zemlak et al. 2009). In a particularly stunning example, Durand et al. (2012) presented evidence for the presence of many cryptic groups within common species of mullets (family Mugilidae), including 13 distinct lineages for the very common *Mugil cephalus;* some of these lineages are known to be reproductively isolated. While this is an exciting field for discoveries, it also means the odds of erroneous identification are much higher in actinopterygian fish than in other vertebrate groups. Discrepancy between studies could be due to using distinct non-identified lineages instead of the same species. Conversely, a single species or population can include individuals with distinct karyotypes (Lautrédou

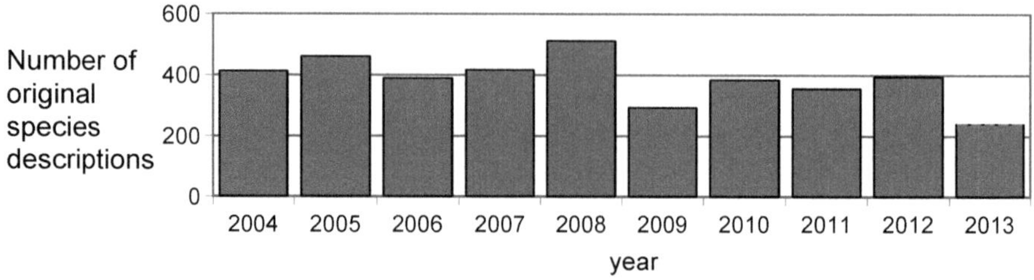

Figure 1. Number of original species descriptions per year.
Source: Fishbase ("List valid species originally described in year" function). Not all species for 2013 are included yet.

et al. 2012). Being able to confirm that they belong to a single evolutionary entity provides for further and more robust discussion for cytogenetic studies.

It is therefore of the uttermost importance for fish cytogenetics studies to have a reliable identification of the specimens, and a way to reassess the identification a posteriori. Identification can be performed using either molecular or morphological approaches (using discrete or continuous characters or character combinations). However these approaches require preserving the whole or part of the studied specimens, along with all the available collection data and must therefore be factored in at the planning stage of the cytogenetics study. Morphological identification can require a lot of expertise, and in recent years molecular identification has been developing at a rapid pace. Collaboration with competent taxonomists working on the same groups ameliorates the identification or taxonomic work, and they might even be grateful for the additional material.

Conservation of some or all of the studied specimens as vouchers in a curated collection permits researchers to check and correct identification at any time, but also to extend and complete the studies to answer previously unforeseen questions or apply other techniques and approaches to complete the study.

There are also practical considerations. Collecting and studying animals is a labour and credit intensive task, with ethical implications. Making the study as complete as possible of each specimen with multiple approaches maximizes the information return per sacrificed animal and capture effort. It can also help to support the collection authorization requests, especially when multiple researchers associate for the different aspects. Lastly, it is not always possible to collect specimens again in case of doubt. Even when it is, all the study has to be repeated on the new specimens, leading to a waste of time and money.

Applications and Limitations

Molecular identification

Molecular identification is based on the comparison of a sequence of the specimen to identify with reference sequences stored in a dedicated sequence library. It has been developing considerably in the last 10 years, especially since the launch of the Barcode of Life project. The standardized marker ("DNA barcode") settled on for the project is a partial mitochondrial gene sequence, the 5′ end of the cytochrome c oxidase subunit I gene (COI). Data description and specimen vouchering are also standardized in the project, and represent a marked amelioration on the low reliability of sequence identification in databases like GenBank. The initiative specifically dedicated to fish, The Fish Barcode of Life (FISH-BOL, http://www.fishbol.org/) now has a reference sequence library including more than 100,000 sequences and 10,000 fish species. DNA barcoding using COI is a reliable approach for the identification of most groups in fish, including Chondrichthyans (Ward et al. 2005; 2009; Dettai et al. 2011). Other commonly used mitochondrial markers present either less variability (12S and 16S rDNAs) or are more difficult to amplify reliably (Cytochrome b and mitochondrial control region). However, species that are very closely related and therefore with little sequence divergence cannot always be identified by this approach. The reliability of the identification also depends largely on the completeness of the reference dataset, and as of now many described species have not yet been sequenced. The Barcode of Life Database (BOLD) is currently the most extensive dataset for any marker.

Many publications describe in detail the protocol for fish DNA barcoding (Ward et al. 2005; 2009; Steinke and Hanner 2011 for instance). It is now relatively cheap (5–10 euros per specimen), and can easily be outsourced to commercial companies. Data analysis is relatively straightforward using the dedicated database (http://www.barcodinglife.org/) with the available tools. Data deposition is similarly simple. The database accepts specimen photographs, generates maps and distance trees and data curation is straightforward. Sequence data for other genes beyond COI can also be deposited, and the sequences transferred very easily to other sequence databases like GenBank before publication.

Importance of a biorepository

Long term conservation of DNA samples or specimens in good conditions is a long standing problem, and requires regular specialised care. Most research institutions do not have curated collections, and precious specimens can end up ruined, lost or even discarded. Curated collections also make the specimens available to a larger scientific community through online databases and loans. Therefore it is best to look up the closest biological repositories with curated collections and contact them prior to the study planning process. The Global Registry of Biorepositories (http://grbio.org/) is an online registry of biological collections in natural history museums and other biorepositories.

Protocols

The preservation protocols are different for molecular and morphological work for the optimal preservation of samples and vouchers. Figure 2 presents the workflow.

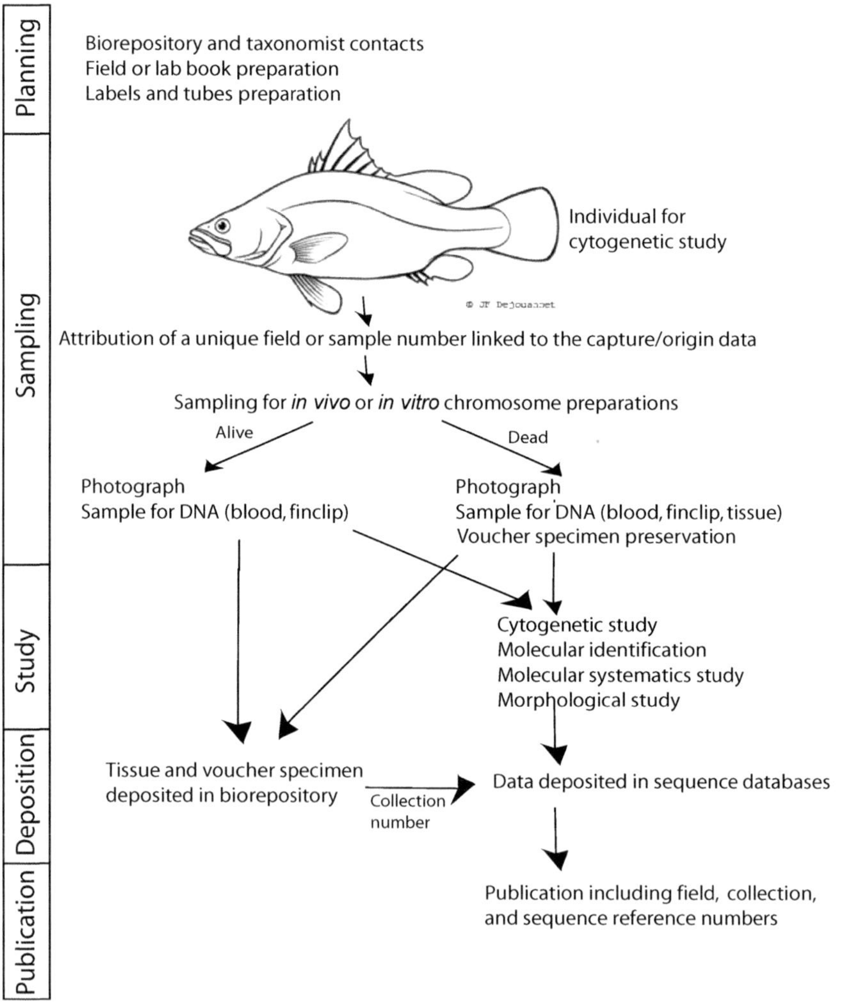

Figure 2. Workflow for the sampling and vouchering process. Fish drawing by J./-F. Dejouannet.

A. Preservation of samples for DNA analysis

DNA is degraded by exposure to a number of chemicals, including formalin. Samples must be taken before preservation with formalin and immediately after sampling for chromosome preparations.

1. Prepare tubes and labels before starting or going to the field. Label the tubes with unique identification numbers, and prepare corresponding labels for the voucher specimens. Double labelling (on the tubes and on labels inside the tubes) is best. Use ethanol resistant collection paper, with pencil inside the tubes and ethanol resistant marker on the outside. 4 or 5 ml tubes are generally the right size. For small specimens, 2 ml tubes may be sufficient.
2. Fill the tubes with the preservative of choice, 95% ethanol or RNA*later* (Ambion). Check the optimal preservation protocol as it can differ for some taxa or environments (Hillyard and Clark 2012). Isopropanol and denaturated ethanol must not be used.
3. Prepare the field notebook in advance indicating all the data categories that need to be completed.
4. In the field: Clean the tools in household concentration unperfumed bleach or through extremely careful ignition after plunging it in concentrated ethanol, then rinse carefully using distilled water and when available, 95% ethanol.
5. Lay the fish on a clean surface, left side down. By convention, the left side should be left untouched, samples are taken from the right side. Specimens should be anesthetized or sacrificed before these steps.
6. A pea-sized (5 mm diameter) muscle piece is enough for most applications. Most common tissue sample types for fish are muscle and fin-clips. Blood can also be used for DNA studies, but needs specific preservatives. Fin-clips and blood can both be used for sampling of fish that will be left alive.
7. Place one sample in each tube. The ethanol to tissue ratio must be three to one or higher, RNA*later* must be 6 to one or higher. The samples will degrade if there is too much tissue relative to the conservation agent.
8. Clean the tools as in 4, between each specimen, or use single-use tools.
9. Leave the samples in RNA*later* in a refrigerator for 24 h so that the preservative penetrates the tissue and then place in a freezer. Refrigerate at +4°C samples in ethanol as soon as possible.
10. Conserve the samples in a freezer.

Small specimens can be preserved directly in ethanol and sampled later for DNA studies, although it is not optimal; formalin provides a faster fixation. The ethanol/tissue ratio must be respected, and the ethanol should be changed after a few hours for better preservation.

B. Specimen photography

This step can be done before or after tissue sampling, on anesthetized or dead specimens. The following description is for a minimal setup (Fig. 3).

Take photographs of all specimens, especially the ones that are released afterwards or where no voucher is kept. Preserved specimen undergo color changes, so taking a picture of the fresh specimen before vouchering is important. For instance, Holm (2011) and Motomura and Ishikawa (2013) review how to take high quality fish specimen pictures in the field. The following description is for a minimal setup.

1. Lay the specimen on a single colored background, left side up (head facing left).
2. Include a size and color scale and the sample identifier number in the picture.
3. Take a high resolution photograph. Check the quality before moving on to the next specimen.

While it is important to respect the standards of the field, it is better to have a photograph not respecting some criteria than having no picture at all.

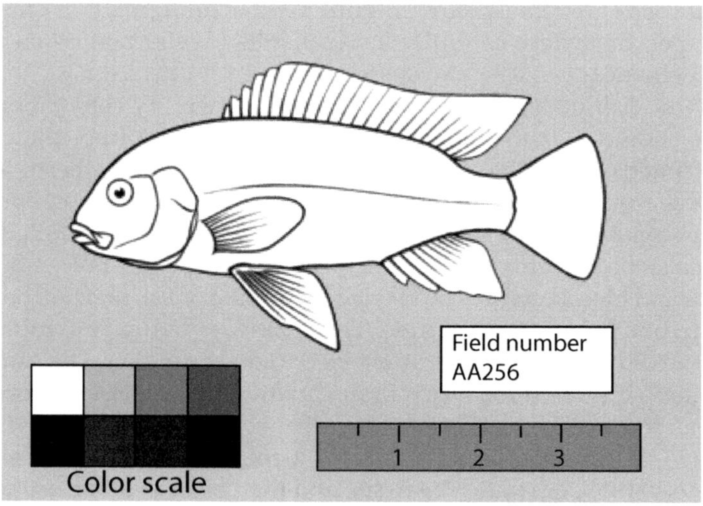

Figure 3. Fish specimen photograph. Fish drawing by J./-F. Dejouannet.

C. Voucher fixation and preservation

Formalin is irritating to skin and airways, carcinogenic and can provoke allergies. All formalin manipulations should be done with adequate protection (adapted filter mask and gloves, or under an adapted aspiration hood and gloves), and the formalin must be disposed of by the adequate chemical waste management channels of the country.

1. Tag or label the specimen with a unique field identifier. The label must be attached in a reliable way to the specimen.
2. Take samples for the DNA analyses, and photographs before preparing the specimen for preservation.
3. When possible, prepare the specimen for fixation. The shape will remain fixed afterwards (see for instance Mayden and Dillman 2008).
4. In a container of a sufficient size to accommodate the full length of the specimen, introduce 10% buffered formalin and the specimen. Do not overfill.
5. Leave the specimen in the formalin for fixation. Fixation time depends on the size of the specimen, from a week for smaller specimens to a month for larger ones.
6. At the end of the fixation time, rinse the specimen with water. Put it in room-temperature water for 24 h. Rinse and repeat to eliminate as much formalin as possible. The formalin-containing water must also be disposed of properly, and manipulated with the proper precautions.
7. Transfer the specimens to 70% ethanol for long term conservation.
8. Transfer the preserved specimens to a curated collection along with their collection data.

Troubleshooting

Number of specimens to preserve

When the individuals are sacrificed anyway, it is desirable to keep as many complete specimens as possible for wild-caught or rare species. For laboratory strains, one or two might be sufficient, but the identification of specimens acquired through the aquarium-trade must be considered with caution, as hybrids, cryptic and undescribed species are regularly sold (Steinke et al. 2009), and there might be unrecognized heterogeneity. When the individuals are sampled and released, the number of voucher specimens depends on the authorizations and local regulations. Moreover, space is limited in most collections. For smaller species or a limited number of specimens, keeping all is desirable and might be feasible. Otherwise, as many as possible should be kept, with a careful selection representative of capture localities, species, populations and morphotypes. In any case, photographs and sampling for molecular studies (see below) do not suffer from these limitations, and should be done on all specimens whether vouchered or not.

Data and labels

The importance of specimen documentation cannot be stressed enough, whatever the type of study, as a specimen loses a lot of its value without it. Both specimen and sequence databases can accomodate the documentation for optimal conservation and availability after the publication. Collection data should be as precise as possible (capture location, depth, collector, collection date, water parameters, etc.) and use predefined fields to ensure consistency across records and avoid forgetting details. Steinke and Hanner (2011) or the BOLD Handbook are good starting points for minimal fields. Data should be transfered to a dedicated database (or at least data file) as soon as possible.

Conserving the link between samples, and between samples and documentation (including photographs) is central for the scientific use of the specimens and samples. It should be carefully planned, with a unique identifier for each sampled individual. The link between the different sample identifiers for the samples originating from the same individual must be registered in the field book, and all samples originating from the same individual can also share a field identifier or part of a field identifier. In all these steps, redundancy of the information between labels and field book helps prevent information losses and to identify possible mistakes. Double labelling (on the tubes and on labels inside the tubes) is helpful for the same reasons.

The link between multiple samples and with the voucher specimen must be conserved and appear clearly in the published manuscript and in the databanks. It is a considerable added value for reliability and scientific precision. More and more journals require for publication that voucher specimens be deposited in a permanent collection, and the collection numbers indicated in the manuscript.

Complete protocol resources

Mayden, R. and C. Dillman. 2008. Tissue and Fish Sample Collection Protocols. http://cypriniformes.org/tissue_fish_%20sample_col_protocols.html Retrieved 02/05/204.

Motomura and Ishikawa. 2013. Fish collection building and procedures manual. English edition The Kagoshima University Museum, Kagoshima and the Research Institute for Humanity and Nature. Kyoto. 70 pp.

Steinke, D. and R. Hanner. 2011. The FISH-BOL collaborators' protocol. Mitochondrial DNA, October 2011; 22(S1): 10–14.

Laboratory set up

Equipment

Major

Fridge (4°C) and Freezer; formalin adapted hood (or filter mask, see below); photographic equipment.

Minor

Formalin-adapted filter mask and formalin-adapted gloves; dissection kit (scalpels, scissors, pliers); beakers; field or laboratory book; jars (for the specimens); ethanol resistant marker and pencil, ethanol resistant paper; polypropylene hermetic tubes.

Biochemicals and reagents

70%, 95% ethanol; 37% formalin solution; household bleach; distilled water; RNA*later* (Ambion).

Stock and working solutions

10% buffered formalin 1 L

100 ml part formalin solution (37%)

900 ml water

3 ml of borax or a few pinches of sodium bicarbonate.

References

Dettai, A., A.-C. Lautredou, C. Bonillo, E. Goimbault, F. Busson, R. Causse, A. Couloux, C. Cruaud, G. Duhamel, G. Denys, M. Hautecoeur, S. Iglesias, P. Koubbi, G. Lecointre, M. Moteki, P. Pruvost, S. Tercerie and C. Ozouf. 2011. The actinopterygian diversity of the CEAMARC cruises: Barcoding and molecular taxonomy as a multi level tool for new findings. Deep Sea Res. II. 58 (1-2): 250–263.

Froese, R. and D. Pauly. 2014. Editors. FishBase. World Wide Web electronic publication. www.fishbase.org, version (05/2014).

Durand, J.D., K.N. Shen, W.J. Chen, B.W. Jamandre, H. Blel, K. Diop, M. Nirchio, F.J. Garcia de León, A.K. Whitfield, C.W. Chang and P. Borsa. 2012. Systematics of the grey mullets (Teleostei: Mugiliformes: Mugilidae): molecular phylogenetic evidence challenges two centuries of morphology-based taxonomy. Mol. Phylogenet. Evol. 64(1): 73–92.

Hillyard, G. and M. Clark. 2012. RNA preservation of Antarctic marine invertebrates. *Polar Biol.* 35(4): 633–636.

Holm, E. 2011. Improved technique for fish specimen photography in the field. Canad. J. Zool. 67(9): 2329–2332.

Lautrédou, A.C., D.D. Hinsinger, C. Gallut, C.H. Cheng, M. Berkani, C. Ozouf-Costaz, C. Cruaud, G. Lecointre and A. Dettai. 2012. Phylogenetic footprints of an Antarctic radiation: the Trematominae (Notothenioidei, Teleostei). Mol. Phylogenet. Evol. 65(1): 87–101.

Mayden, R. and C. Dillman. 2008. Tissue and Fish Sample Collection Protocols. http://cypriniformes.org/tissue_fish_%20sample_col_protocols.html Retrieved 02/05/204.

Motomura, H. and S. Ishikawa. 2013. Fish collection building and procedures manual. English edition. The Kagoshima University Museum, Kagoshima and the Research Institute for Humanity and Nature, Kyoto. 70 pp.

Thomas, R.C., Jr., D.A. Willette, K.E. Carpenter and M.D. Santos. 2014. Hidden diversity in sardines: genetic and morphological evidence for cryptic species in the goldstripe sardinella, Sardinella gibbosa (Bleeker, 1849). PLoS One 9(1): e84719.

Steinke, D., T.S. Zemlak and P.D.N. Hebert. 2009. Barcoding nemo: DNA-based identifications for the ornamental fish trade. PLoS One 4(7): e6300.

Steinke, D. and R. Hanner. 2011. The FISH-BOL collaborators' protocol. Mitochondrial DNA 22(S1): 10–14.

Ward, R.D., T.S. Zemlak, B.H. Innes, P.R. Last and P.D.N. Hebert. 2005. DNA barcoding Australia's fish species. Philos. Trans. R. Soc. B Biol. Sci. 360: 1847–1857.

Ward, R.D., R. Hanner and P.D.N. Hebert. 2009. The campaign to DNA barcode all fishes, FISH-BOL. J. Fish Biol. 74(2): 329–356.

Zemlak, T.S., R.D. Ward, A.D. Conell, B.H. Holmes and P.D.N. Hebert. 2009. DNA barcoding reveals overlooked marine fishes. Mol. Ecol. Res. 9: 237–242.

3

Direct Chromosome Preparation from Freshwater Teleost Fishes

L.A.C. Bertollo,[a,*] *M.B. Cioffi*[b] and *O. Moreira-Filho*[c]

[1]Universidade Federal de São Carlos, Departamento de Genética e Evolução, Caixa Postal 676, 13565-905 São Carlos, SP, Brazil.
[a]e-mail: bertollo@ufscar.br
[b]e-mail: mbcioffi@ufscar.br
[c]e-mail: omfilho@ufscar.br
*Corresponding author

Principles

In recent years, cytogenetics has contributed to a significant development in fish research. This has certainly been due to the improvement of methodological approaches that allowed a better characterization of the chromosomes, with implication for the comprehension of fish karyotypic diversity and evolution. Certainly a good chromosome preparation represents the basic step for any conventional or molecular cytogenetic protocol.

Different tissues may be used as a source of mitotic cells such as kidney, spleen and gills. In teleost fishes, the "head" kidney (most cranial portion) contains predominantly hematopoietic and lymphoid tissue (Hybiya 1982), thus considered more advantageous in terms of metaphase quantity and quality.

Although the shape of the kidney may vary among species, in teleost fishes it is clearly subdivided in an anterior portion close to the skull (head kidney) and a posterior portion that extends along the body towards the tail (body kidney). The head kidney is derived from the pronephros and the body kidney from the mesonephros (Hybiya 1982). In comparison, the chromosome preparations obtained from the anterior kidney are free of tissue fibers that are frequently observed in preparations from its posterior region.

Applications and Limitations

The technique here described may be used for directly obtaining chromosomes from various teleost fish tissues. The entire or part of an organ may be used as a source of mitotic cells depending on the size of the specimen. Nonetheless, the kidney cells, especially from the head kidney, provide the best preparations, being used almost exclusively in our fish cytogenetics studies (Fig. 1).

In specimens of medium to large sizes, the extraction of the head kidney is facilitated by its relatively large size. For small-sized specimens the procedure described in Appendix 1 facilitates the exposure and removal of this organ (Moreira-Filho and Bertollo 1991).

A variation of the present methodology is presented in Appendix 2, which can be used for obtaining mitotic chromosome spreads in field conditions (Blanco et al. 2012).

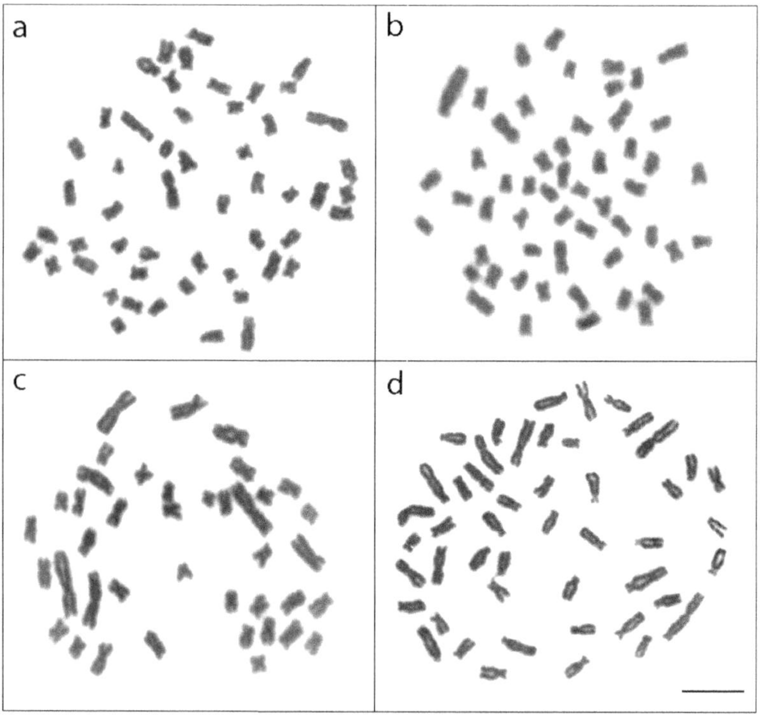

Figure 1. Giemsa-stained metaphase plates of (a) *Triportheus auritus*, 2n = 52 (Characiformes, Characidae); (b) *Leporinus elongatus*, 2n = 54 (Characiformes, Anostomidae); (c) *Hoplias malabaricus*, 2n = 40 (Characiformes, Erythrinidae); (d) *Astyanax scabripinnis*, 2n = 48 (Characiformes, Characidae) obtained from kidney cells by direct preparation. Bar = 5 μm.

Protocol

This protocol is adapted to teleosts, and the temperature and time of hypotonic treatment are especially suitable for Neotropical freshwater fishes. Thus, some adjustments, such as temperature, time of treatment with colchicine and hypotonization, are probably needed for other fish groups living in different environmental conditions.

1. Inject an aqueous colchicine solution (0.025%) into the abdominal cavity of the specimen, between the pectoral and pelvic fins (1 ml/100 g body weight). Take care that the injection is superficial in order to avoid lethal injuries to internal organs such as to the swim bladder or other vital organs;
2. Keep the fish in an aerated aquarium for 45–50 minutes, and then sacrifice it after anaesthetization through immersion in diluted benzocaine solution;
3. Place the animal in a ventral position and open the abdomen from the anal opening to the base of the isthmus, exposing the anterior kidney. Remove approximately 1.0–1.5 cm of the kidney tissue (for small-sized specimens, follow steps described in the appendix);
4. Quickly wash the pieces of kidney in distilled water or hypotonic solution (0.075 M KCl) and transfer the material to a small Petri dish containing approximately 10 ml of hypotonic solution;
5. Dissociate the tissue with the aid of dissection tweezers, completing this process with a needle-free hypodermal syringe through delicate aspiration and expiration of the cells to facilitate their separation, until a homogeneous cell suspension is obtained;
6. Incubate the cell suspension at 36–37°C in an incubator for 20 minutes;
7. Carefully mix well the cell suspension with a Pasteur pipette and transfer it to a centrifuge tube, discarding larger tissue fragments;
8. Add a few drops (5–10) of freshly prepared Carnoy's fixative, mixing well with the cell suspension and centrifuge for 10 minutes at 500–800 rpm;
9. Discard the supernatant with a Pasteur pipette and add 5–7 ml of fixative, letting it slide along the centrifuge tube walls;
10. Carefully mix the cell pellet and the fixative, with the aid of a Pasteur pipette;
11. Repeat items 7 to 9 two more times;
12. After the last centrifugation and supernatant elimination, add approximately 1 ml of fixative and mix the pellet well, obtaining a cellular suspension neither too concentrated nor too diluted (The quantity of added fixative will vary, depending on the quantity of the cell pellet obtained);
13. Store the cell suspension in "Eppendorf" tube, which may be kept frozen or proceed directly to the slide preparations according to the following steps:

14. Using a Pasteur pipette release 3 to 4 drops of cell suspension upon different regions of a cold and well-cleaned slide previously maintained in distilled water in the refrigerator. Good results may also be obtained if the slide is heated close to 25°C–30°C, keeping it on a hot plate;
15. Drain the excess material or water (in the case of humid slides) by slightly tilting the slide on filter paper;
16. Air-dry the preparations and stain with 5% Giemsa in a phosphate buffer, pH 6.8, for 5 to 8 minutes;
17. Wash the slide in running water and air-dry.

Troubleshooting

1. Sampling head kidney tissue from small fishes

The following steps, illustrated in Fig. 2, facilitate the exposure and removal of head kidney in small-sized specimens (from Moreira-Filho and Bertollo 1991).

1. Position the fish with the ventral surface faced up, make a transversal cut at the base of the isthmus (A), followed by a longitudinal opening reaching the anus (B).
2. Separate the laterally bisected parts, pinning them upon a base to expose the digestive tract (C).

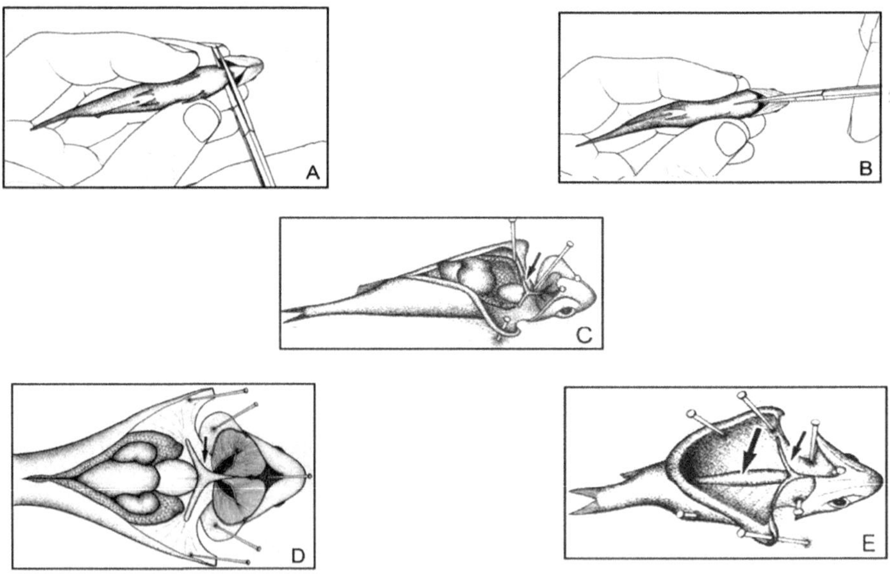

Figure 2. Exposure and extraction of the anterior and/or posterior kidney in a small-sized fish. The small arrows indicate the anterior (head) kidney and the large arrows the posterior (body) kidney (reorganized from Moreira-Filho and Bertollo 1991).

3. Cut the region between the pharynx and the esophagus, moving the viscera to the posterior region of the body with dissection tweezers to expose the anterior kidney (D).
4. Finish the gut removal, also exposing the posterior kidney (E).
5. Remove the kidney tissue with dissection tweezers.

2. Obtaining mitotic chromosome spreads in field conditions

Good chromosomal preparations can be also obtained with the adaptation of this methodology for field conditions, when collection sites are far from research laboratories or even when working with highly sensitive species (Blanco et al. 2012).

1. Submit the specimens to conventional colchicine treatment (intraperitoneal injection of 0.025% solution, 1 ml per 100 g body mass), keeping them alive for 30–50 min;
2. Remove approximately 1.0–1.5 cm of the kidney tissues and wash the pieces in distilled water or hypotonic solution (0.075 M KCl);
3. Dissociate the tissues in 10 ml of hypotonic solution (0.075 M KCl) using a glass syringe without a needle until complete homogenization of the material;
4. Keep the cell suspension at room temperature (26°C–28°C) for 30 min;
5. Add 2 ml of fixative solution (3:1 methanol : glacial acetic acid) and homogenize the material with a Pasteur pipette;
6. Store the material in a centrifuge tube and keep it sealed at room temperature (26°C–28°C) or under refrigeration (–20°C) up to seven days (168 h tested) before further processing in the laboratory;
7. Homogenize the material with a Pasteur pipette and centrifuge for 10 minutes at 500–800 rpm;
8. Continue processing the material according to steps 9–17 of the above protocol.

Laboratory set up

Equipment

Major

Aerated aquarium; analytical balance; common balance; centrifuge; hot plate; water distiller; incubator; freezer; pHmeter; refrigerator.

Minor

Pins; hydrophylic cotton; glass chamber; "Eppendorf" tubes; Microscopy slides; filter paper; dissection tweezers (fine tips); dissection plates; small

Petri dishes (20–30 ml); graduated cylinders; hypodermal syringe with needle; dissection scissors (fine tips); timer; centrifuge tubes (15 ml).

Biochemical and reagents

Glacial acetic acid; distilled water; benzocaine ($NH_2C_6H_4COOC_2H_5$; Ethyl p-aminobenzoate; ethyl PABA; anesthetic); colchicine; Giemsa stain; KCl; KH_2PO_4; absolute methanol (PA); absolute ethanol; NaOH.

Stock and working solutions

Benzocaine Solution

0.5 g of benzocaine
5 ml of absolute ethanol
5.000 ml of water
Dissolve the benzocaine in the ethanol and add the water.

Colchicine solution

0.025 g of colchicine
100 ml of distilled (or deionized) water.

References

Blanco, D.R., L.A.C. Bertollo, R.L. Lui, M.R. Vicari, V.P. Margarido, R.F. Artoni and O. Moreira-Filho. 2012. A new technique for obtaining mitotic chromosome spreads from fishes in the field. J. Fish Biol. 81: 351–357.

Hybiya, T. 1982. Kidney. pp. 94–104. *In*: Takashima, F. and T. Hybiya (eds.). An Atlas of Fish Histology—Normal and Pathological Features. Gustaw Fisher Verlog, Tokyo.

Moreira-Filho, O. and L.A.C. Bertollo. 1991. Extraction and use of the cephalic kidney for chromosome studies in small fish. Braz. J. Genet. 14: 1085–1090.

4

Direct Mitotic Chromosome Preparations from Chondrichthyan Tissues

L. Rocco

University of Napoli 2, Department of Environmental, Biological and Pharmaceutical Sciences and Technologies, Via Vivaldi 43, 81100, Caserta, Italy.
e-mail: lucia.rocco@unina2.it

Principles

The karyotype of the Chondrichthyes is still not well known, principally due to the practical problems such as the body size and life style of many species that prevent keeping them alive. Moreover they often have a huge number of chromosomes, making the karyotype study difficult.

There is no reason to infer any property of their chromosomal organization from what is known from other "fishes": chondrichthyans have their own phylogenetic lineage separated from osteichthyans since c.a. 420 million years ago.

Applications and Limitations

Up to about 15 years ago the papers present in literature on the karyotype of the Chondrichthyes used a great number of chromosomal plates that often did not show chromosomes of good morphological quality. It thus followed that often there was no agreement on the fundamental number (FN) or even on the chromosome number (see Stingo and Rocco 2001 for a review).

Technical problems remained, such as variable quality of the chromosomal morphology in relation to the level of condensation, due to the use of different techniques and the inevitable margin of arbitrariness in the classification of the chromosomes.

The technique that is described here gives the possibility of obtaining many well-prepared chromosomes from both young and adult specimens of Chondrichthyes. This is the starting point to acquire further information on some important cytogenetic characteristics such as constitutive heterochromatin and the nucleolar organizing regions (NORs), by means of banding techniques (Rocco et al. 2002), or the chromosomal localization of sequences of DNA having particular structural or functional meaning (Rocco 2013).

The methodology described in this chapter follows the procedure of "air-drying" from somatic tissue, modified for chondrichthyan chromosomes. These fish have a much slower metabolism than the teleosts, participating in osmoregulation with a high percentage of urea (350–600 mM) in the blood (Robertson 1989; Tam et al. 2003). For this reason reaction times to colchicine treatment is longer than those normally used for other *in vivo* treatment protocols. Furthermore, the hypotonic treatment generally used for the other vertebrates (0.075 M KCl) does not produce a sufficient number of analyzable chromosome spreads in the preparations.

Protocol

a) After being anaesthetized, each animal receives an injection of 0.1 ml colchicine (0.5 mg/ml) per 10 g body weight.
b) The colchicine is injected into the abdominal cavity and left to act for about 16 h before sacrifice and dissection. In cartilaginous fish the highest mitotic index is found in the spleen, the gills and the gonads. The testes often provide good meiotic chromosome plates.
c) The tissues are incubated in a hypotonic solution containing 0.3% KCl, 0.25% Na Citrate and 0.28% NaCl for 45 min at room temperature.
   The treatment with the hypotonic solution allows the cells to swell, for better chromosome spreading.
d) At the end of the hypotonic treatment, the tissue samples are sieved through a 100 μm mesh sieve and then prefixed in conventional fixative.
e) The pre-fixed cells are then centrifuged for 10 min at 1000 rpm. Freshly prepared fixative is changed thrice.
f) The cell pellet is finally re-suspended in about 500–1000 μl of fresh fixative.
g) Glass slides are washed and rinsed in triple distilled water, then kept in ice-cold water prior to preparing the chromosome spreads.
h) The cell suspension (100 μl) is dropped onto the wet, ice-cold slide from the height of 30–40 cm. Drop the suspension slowly onto the slide, 1 drop at a time, moving the pipetman so that single drops fall on the surface area without overlapping. The fixative quickly evaporates from under and around the cells and the cells flatten completely, so that the chromosomes spread out.

i) The quality of the slide preparation is checked with a contrast phase microscope. If cytoplasm is detected, it is dispersed by washing the cells several more times with fresh fixative before preparing more slides. If the suspension is too concentrated or too diluted this should be corrected by adding or removing fixative. Concentrated cell suspensions will produce under-spread metaphases, while it is time consuming to search for metaphase plates in too diluted suspensions.
j) The slides are air dried vertically and transferred to a slide box. Parafilm is wrapped around the box prior to freezing the slides at –20° or (better) at –80°C. According to our experience, slides can be stored for 1 month at –20°C or for 1 year at –80°C.
k) To check chromosome morphology and for establishing the karyotype, Giemsa staining is used. In the example provided here (Fig. 1) the

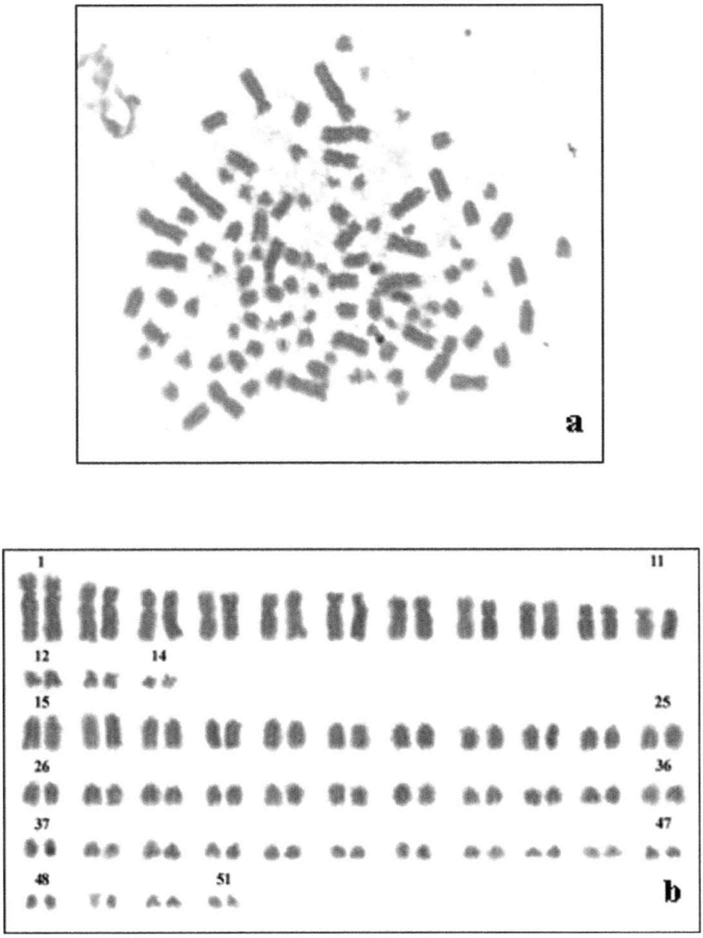

Figure 1. Metaphase plate (a) and karyotype (b) of *Ginglymostoma cirratum* (Chondrichthyes, Ginglymostomatidae), 2n = 102.

slides were stained in a Hellendhal jar for 5 min in a 5% Giemsa in McIlvaine's buffer (pH 7), then well rinsed in distilled water and air dried. The chromosome plates were observed with a high resolution light photomicroscope with an oil immersion objective 100X PlanApo, the metaphases were recorded on a CCD camera and analyzed by a software for the semi-automatic reconstruction of the karyotype.

Troubleshooting

The room temperature and the humidity will affect the drying time of cells and the quality of the metaphases. It is essential to try to control all the variables to achieve the optimum conditions. The use of vaporizers when the air is extremely dry and the use of air conditioners when the room temperature is above 22°C will help to produce high quality metaphases. Our experimented optimum conditions are: room temperature between 20°C and 22°C and humidity between 45 and 50%.

We do not store slides at +4°C, because the quality of the chromosomes could deteriorate.

Laboratory set up

Equipment

Major

Precision balance; pHmeter; low speed centrifuge with mobile rotor.

Minor

15 ml conical centrifuge tubes; graduated cylinders; glass Hellendhal jars with lids; Pasteur pipettes; bulbs for the Pasteur pipettes; superfrost slides; sterile 1 or 5 ml syringe; timer; mesh sieve.

Biochemicals and reagents

KARYOMAX Stock Solution (LIFE TECHNOLOGIES, 10092-013); potassium chloride (CARLO ERBA, 471177); sodium chloride (APPLICHEM, A2942); tri-Sodium citrate dihydrate (CARLO ERBA, 479487); glacial acetic acid (CARLO ERBA, 401391); methanol (CARLO ERBA, 414816); colchicine (SIGMA, C-9754); anhydrous citric acid (SIGMA, C2404); anhydrous dibasic sodium phosphate (FLUKA, 71639).

References

Robertson, J.D. 1989. Osmotic constituents of the blood plasma and parietal muscle of *Scyliorhinus canicula* (L.). Comp. Biochem. Physiol. 93A: 799–805.

Rocco, L. 2013. Sex-related genomic sequences in cartilaginous fish: an overview. Cytogenet. Genome Res. 141: 169–176.

Rocco, L., M.A. Morescalchi, D. Costagliola and V. Stingo. 2002. Karyotype and genome characterization in four cartilaginous fishes. Gene. 295: 289–298.

Stingo, V. and L. Rocco. 2001. Selachian cytogenetics: a review. Genetica. 111: 329–347.

Tam, W.L., W.P. Wong, A.M. Loong, K.C. Hiong, S.F. Chew, J.S. Ballantyne and Y.K. Ip. 2003. The osmotic response of the Asian freshwater stingray (*Himantura signifer*) to increased salinity: a comparison with marine (*Taeniura lymma*) and Amazonian freshwater (*Potamotrygon motoro*) stingrays. J. Exp. Biol. 206: 2931–2940.

5

Mitotic Chromosome Preparations of Freshwater Stingrays

*V. Paes da Cruz** and *F. Foresti*

Departamento de Morfologia, Instituto de Biociências de Botucatu, UNESP, Botucatu – SP, Brasil.
e-mail: cruzvp@outlook.com
*Corresponding author

Principles

The technique for obtaining mitotic chromosomes proposed by Foresti et al. (1981) when applied in freshwater stingrays didn't result in good quality chromosome preparations. Only a few studies have used this technique to obtain good quality chromosomes in this fish group (Valentim et al. 2006; Cruz et al. 2011). Therefore, it was necessary to adapt the technique to freshwater stingrays. Classical *in vivo* methods basically consist of mitotic stimulation followed by colchicine injection, according to the technique described in Chapter 2. In this chapter we present modifications of this technique to obtain a better quality of chromosomes in stingrays. Typically, the kidney tissue is used in this technique, but in stingrays the tissue that showed the best quality and quantity of metaphases was the spleen. Other important modifications are the time of colchicine and hypotonic treatments.

Applications and Limitations

The air-drying technique has been widely used in fishes which involves pretreatment with colchicine (Foresti et al. 1981). The procedure is usually employed after live specimens have been transported to the laboratory or another location that offers the necessary infrastructure to keep them alive. The requirement for laboratory facilities has therefore been a limiting factor in the study of many fish species, especially those that require controlled temperature and oxygen conditions for survival or are collected far from the laboratory such as the stingrays.

Besides these difficulties, the classical technique used in fish cytogenetics is probably not successful in stingrays due to the particular biological characteristics of these animals. The ancestor of stingrays, the marine elasmobranchs developed specific strategies to withstand salinity which involve low metabolism and the retention of high levels of urea (Speers-Roesch et al. 2006). However, freshwater elasmobranchs possess low levels of urea as a result of evolutionary adaptation to the hypotonic environment, but still have low metabolism, which could slow down the process of mitotic division and possibly affect colchicine treatment.

Protocol

1. Colchicine treatment

a) Inject 0.05% colchicine, at a ratio of 0.5 ml/100 g of body weight into the abdominal cavity.
b) Maintain the fish in a well-aerated tank for a period of four hours.

2. Obtention of mitotic chromosomes (modified from Foresti et al. 1981)

a) Anesthetize the stingrays in a water bath containing 10% Benzocaine. Follow Fig. 1.
b) Open the animal cavity on the ventral side (Fig. 1b), remove the liver (Fig. 1c); it is extremely large and may hide the spleen. After removal of the liver (Fig. 1d), the intestine can be visualized, and the spleen can be seen adhered to the intestine (Fig. 1e).
c) Remove the spleen and clean it with distilled water to remove blood and other fluids. Cut and remove a small piece of tissue in the central region of this organ (Fig. 1f).
d) Place the tissue in a Petri dish and add 7 ml of prewarmed (37°C) hypotonic solution (0.075 M KCl).
e) Homogenize cell suspension by teasing tissue with fine-tipped tweezers, and then homogenize by pumping with a Pasteur pipette.
f) Transfer the suspension into a 15 ml centrifuge tube, and incubate at 37°C for 30 minutes.
g) Prefix cells by adding 5 to 7 drops of freshly prepared ice cold fixative in the tube, and gently mix the suspension with the Pasteur pipette. Centrifuge at 1000 RPM for 10 min.

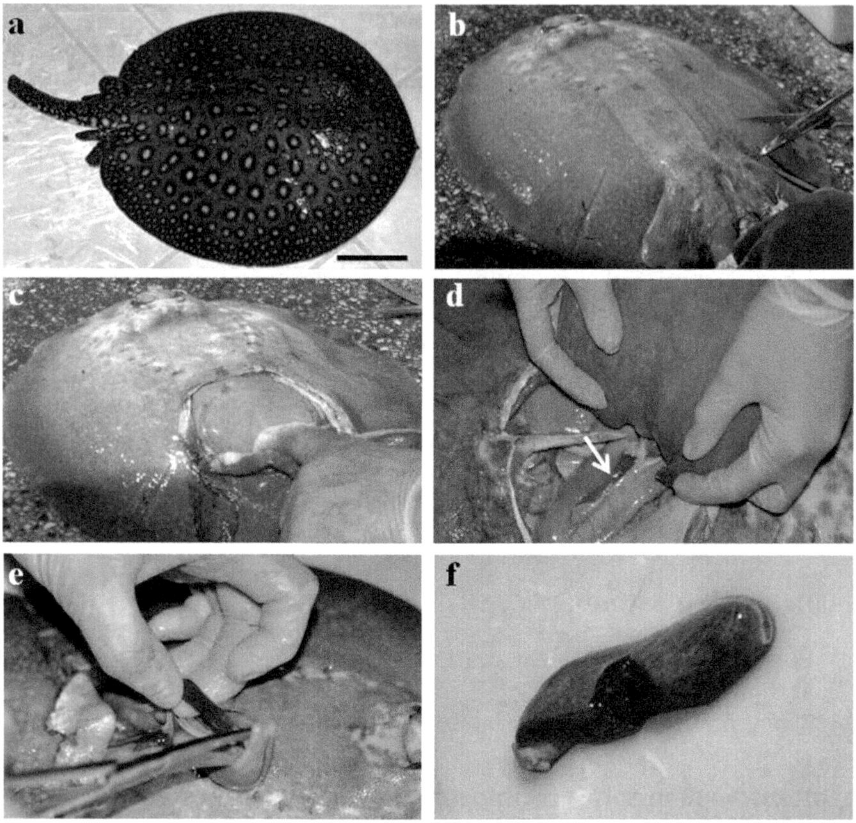

Figure 1. Stingray tissue sampling for chromosome preparations. (a) *Potamotrygon motoro* (male); (b) ventral view of the body; (c) abdominal cavity showing the liver (yellow); (d) removing the liver, organs are displayed; the arrow indicates the spleen, adhered to the intestine; (e) removal of the spleen, and (f) cut the central region of the spleen.

Color image of this figure appears in the color plate section at the end of the book.

h) Discard the supernatant, and progressively carefully resuspend the pellet by adding 7 ml of freshly prepared fixative, to avoid cell aggregates. Centrifuge for 10 minutes at 1000 RPM.
i) Discard the supernatant and resuspend the cell pellet with fresh fixative. Centrifuge and fix two more times.
j) After the final centrifugation, homogenize the pellet in 1 ml of fresh fixative. Either drop cell suspension on the slides or store in an Eppendorf tube in the deep-freezer (–20°C).

3. Chromosomal analysis

Prior to chromosomal spreading gently clean microscope slides by washing them in 99% ethanol: 1% concentrated HCI solution. Swipe with optic tissue and dry. Using a Pasteur pipette drop cell suspension (1 drop per slide) on the microscope slide from a height about of 20 cm and dry slides in flat position. After air-drying, the chromosomal preparations can be Giemsa stained for 10 min, rinsed in running water and air-dried (see as example, Fig. 2).

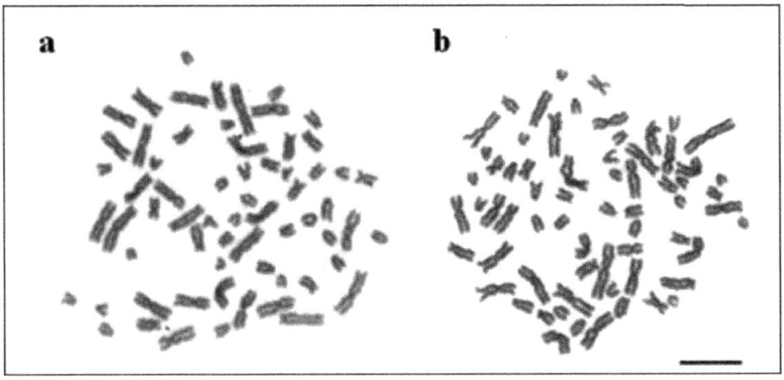

Figure 2. Metaphase plates of *Potamotrygon motoro*; (a) male and (b) female.

Troubleshooting

Manipulation of animals: Stingrays are venomous and may cause serious injuries with their tail stingers. Care should be taken when manipulating these animals.

Storage tank: Due to the size of individuals which can vary from small to large they must be stored in large water tanks. It is important to control the number of individuals per tank because they release large concentrations of urea in the water, which is toxic for the animals.

Tissue for chromosome preparation: Avoid removing the ends of the spleen instead the mid part, because the ends are full of membranes and difficult to dissociate.

Laboratory Set up

Equipment

Autoclave—not necessary if plastic disposable material is used; small bench centrifuge for speeds up to 1000 RPM; centrifuge conical tubes—approximate capacity 15 ml; 37°C incubator or water bath; microscope; 1 ml and 20 ml sterile syringes and needles; sterile 10 ml Pasteur pipettes; bulbs for the Pasteur pipettes; microscope slides; pHmeter, glass Hellendhal jars with lids; Eppendorf tubes; timer, dissecting tools.

Biochemicals and reagents

Giemsa stain stock solution (INVITROGEN, 10092-054); 5% Giemsa in phosphate buffer (pH 6·8); 0.075 M KCL (hypotonic solution); 10% benzocaine.

Stock and working solutions

Colchicine

Stock solution 1 mg/ml stored at –20°C
Working solution 0.05% colchicine.

References

Cruz, V.P., C. Shimabukuro-Dias, C. Oliveira and F. Foresti. 2011. Karyotype description and evidence of multiple sex chromosome system $X_1X_1X_2X_2/X_1X_2Y$ in *Potamotrygon* aff. *motoro* and *P. falkneri* (Chondrichthyes: Potamotrygonidae) in the upper Paraná River basin, Brazil. Neotrop. Ichthyol. 9(1): 201–208.

Foresti, F., L.F. Almeida-Toledo and S.A. Toledo-Filho. 1981. Polymorphic nature of nucleous organizer regions in fishes. Cytogenet. Cell Genet. 31: 137–144.

Oliveira, C., L.F. Almeida-Toledo, F. Foresti and S.A. Toledo-Filho. 1988. Supernumerary chromosomes, Robertsonian rearrangements and multiple NORs in *Corydorasaeneus* (Pisces, Siluriformes, Callichthyidae). Caryologia 41: 227–236.

Speers-Roesch, B., Y.K. Ip and J.S. Ballantyne. 2006. Metabolic organization of freshwater, euryaline, and marine elasmobranchs: implications for the evolution of energy metabolism in sharks and rays. J. Exp. Biol. 209: 2495–2508.

Valentim, F.C.S., J.N. Falcão, J.I.R. Porto and E. Feldberg. 2006. Chromosomes of three freshwater stingrays (Rajiformes, Potamotrygonidae) from the Rio NegroBasin, Amazon, Brazil. Genetica 128: 33–39.

6

Direct Chromosome Preparation from Regenerating Fish Fin Tissue

M. Völker[1] and *P. Ráb*[2,*]

[1]Institut für Evolutionsbiologie und Ökologie, Universität Bonn, 53121 Bonn, Germany.
e-mail: mv53121@googlemail.com

[2]Laboratory of Fish Genetics, Institute of Animal Physiology and Genetics, Academy of Sciences of the Czech Republic, 277 21 Liběchov, Czech Republic.
e-mail: rab@iapg.cas.cz

*Corresponding author

Principles

This protocol for direct chromosome preparation from regenerating fin tissue is based on the methodology developed by Denton and Howell (1969), Scheel (1972), Kligerman and Bloom (1977), with some modifications as described by Völker et al. (2006; 2007). In brief, regenerating fin tissue is incubated in a diluted Ringer solution containing colchicine, which results in combined and simultaneous mitostatic and hypotonic treatments. The piece of fin is then fixed and the tissue minced in aqueous acetic acid to obtain a cell suspension which is dropped onto clean, pre-warmed slides.

In our experience the incubation protocol of fin biopsies here described provides more metaphases plates suitable for cytogenetic analysis than other protocols based on the direct incubation of live fish in a weakly concentrated colchicine solution, as proposed by Cattin and Ferreira (1989).

Applications and Limitations

Chromosome preparation from regenerating fin tissue was the standard method used by Scheel (e.g., 1968; 1972; 1974; 1990) for the cytogenetic analysis of various species of African and Asian killifishes as well as some atheriniform fishes. We successfully applied a modified version of his protocol to *Chromaphyosemion* killifishes (e.g., Völker et al. 2006; 2007; Fig. 1a) and the vaillantellid fish *Vaillantella* (Bohlen et al. 2008), *Danio* (unpublished), *Boraras* (present work) and *Carassius* (Ogawa et al. 1973) (Fig. 1b, c). Provided that the researcher has sufficient preparation skills, the technique described here can be used to study very small fishes (adult specimens of *Boraras maculatus* reach a total length of about 20 mm).

Compared to most other methods of direct chromosome preparation, using regenerating fin tissue as a source of metaphase chromosomes has the advantage that the animals under study do not have to be sacrificed. This is especially important if the specimens are rare, hard to obtain or if they are needed for additional examinations. Some methods for chromosome preparation from cultured cells (e.g., fibroblast cultures) have the same advantage, but they are much more complicated, expensive and time-consuming. The amount

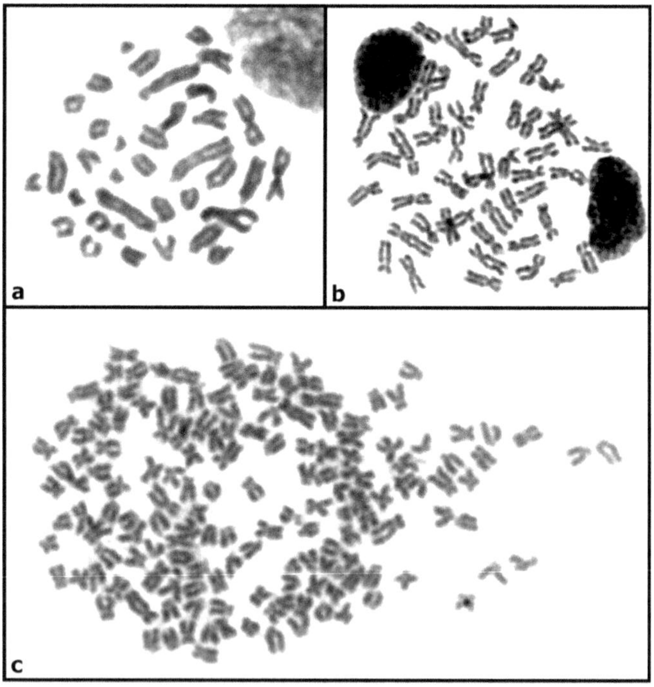

Figure 1. Giemsa-stained metaphase chromosomes from regenerating fin tissue of (a) *Chromaphyosemion* sp. "Rio Muni" GEMHS00/41 (2n = 36), (b) *Boraras maculatus* aquarium stock (2n = 50) and (c) *Carassius gibelio* (3n = 156).

of metaphases that can be obtained by direct chromosome preparation from regenerating fin tissue is lower than in well-established cell cultures, but sufficient to allow for the analysis of highly complicated cases of intraspecific karyotypic variability (e.g., Völker et al. 2006; 2007). Chromosomes prepared according to the protocol described in this chapter can be used for Giemsa staining as well as banding techniques including conventional, $AgNO_3$-staining, C-banding, fluorescence banding with CMA_3 and DAPI (Völker et al. 2006; 2007) as well as Fluorescence *In Situ* Hybridization (FISH).

Protocol

Modified from Denton and Howell (1969), Scheel (1972), Kligerman and Bloom (1977) and Völker et al. (2007).

1. Cut off the hindmost part of the caudal fin from the studied specimen with fine scissors and discard it (a part can be preserved in ethanol 70% as voucher for molecular analyses). Keep the fish under optimum conditions.
2. After three to four days, cut off the regenerated piece of fin with fine scissors and transfer it to a small plastic Petri dish filled with 5 ml of Ringer solution diluted 1:4 with distilled water and containing a final concentration of 0.05% colchicine.
   Incubate for 2 h at the same temperature used for keeping the fish in aquarium.
3. Using a Pasteur pipette, replace the Ringer solution with 5 ml of freshly prepared, cold fixative (3:1 methanol : acetic acid) by gentle pipetting. Incubate at 4°C for 20 min.
4. Change the fixative at least twice and extend the last fixation step overnight.
5. Depending on the size of the regenerated tissue, place the whole piece or part of it on a SIGMA S 0770 mesh whose edges have been folded down and which is placed in a small plastic Petri dish.
6. Using a micropipette, add 40 µl of 50% aqueous acetic acid and mince the tissue very gently for about 10 s using fine, curved forceps.
7. Filter the cell suspension through the mesh and transfer it to a 1.5 ml microtube.
8. Repeat steps 6 and 7 two to three times.
9. Using a micropipette, place about 45 µl of the cell suspension onto a clean, dry slide that has been heated to about 45°C on a slide warmer. After 20 s, suck back the drop into the micropipette, leaving a ring of cells on the slide. Repeat this process twice per slide, creating three adjacent rings of cells.

Troubleshooting

Like with most other methods for chromosome preparation, problems related to the amount and spreading of metaphases may occur. If the slides do not contain a sufficient amount of metaphases, make sure that the fish is kept under optimum conditions. The water temperature in particular has a strong effect on the mitotic index in the regenerating fin tissue. Besides, if no regenerated tissue is visible after three to four days (regenerated fin tissue is colourless and slightly translucent), the time between the biopsy and the sampling of the regenerated tissue may be extended. If visible regeneration has occurred, but the slides still contain only few and badly spread metaphases, the most likely cause is that the mincing of the tissue in acetic acid has been too severe and/or too long. A very gentle treatment of the tissue is essential in obtaining many well-spread metaphases.

If the mitostatic treatment is too strong, resulting in highly contracted chromosomes with separated chromatids, the colchicine concentration should be reduced and/or the duration of the mitostatic treatment shortened.

If the spreading of chromosomes is improper, the duration of the combined mitostatic and hypotonic treatment may be extended and/or a higher dilution of the Ringer solution used. Besides, increasing the temperature of the slide on which the cell suspension is dropped, increasing the size of the rings of cells on the slide and decreasing the time that the suspension remains on the slide may improve spreading (Kligerman and Bloom 1977). Since a good spreading of chromosomes requires the use of a sufficient amount of hypotonic medium (Foresti et al. 1992), the amount of diluted Ringer solution may have to be increased when working with large biopsies.

Finally, the Ringer solution given in this protocol is designed for freshwater teleosts and may be substituted for an appropriate Ringer solution if the method is applied to marine species.

Laboratory set up

Equipment

Major

Slide warmer, adjustable to 45°C; refrigerator (4°C).

Minor

Fine scissors; plastic petri dishes (diameter 55 mm); fine curved forceps; Pasteur pipettes; SIGMA S 0770 m cell dissociation sieve, mesh size 380 µm; Micropipette (10–100 µl) with tips; 1.5 ml microtubes; clean, dry microscope slides; timer.

Biochemicals and reagents

$CaCl_2$; KCl; NaCl; $NaHCO_3$; colchicine; methanol; glacial acetic acid.

Stock solutions

Ringer solution for freshwater teleosts, 10x (Ogawa et al. 1973):

1.281 M NaCl

0.025 M KCl

0.018 M $CaCl_2$

0.002 M $NaHCO_3$

Dilute 1:40 with distilled water and add colchicine to a final concentration of 0.05%.

References

Bohlen, J., M. Völker, M. Rábová and P. Ráb. 2008. Note on the banded karyotype of the enigmatic South Asian loach fish *Vaillantella maassi* (Cypriniformes). Ichthyol. Res. 55(1): 82–84.

Cattin, P.M. and J.T. Ferreira. 1989. A rapid, non-sacrificial chromosome preparation technique for freshwater teleosts. S. Afr. J. Zool. 24: 76–78.

Denton, T.E. and W.M. Howell. 1969. A technique for obtaining chromosomes from the scale epithelium of teleost fishes. Copeia 1969: 392–393.

Foresti, F., C. Ozouf-Costaz and P. Ráb. 1992. Volume dependent effect of hypotonic treatment on good spreading of fish chromosomes. Cybium 16: 209–212.

Kligerman, A.D. and S.E. Bloom. 1977. Rapid chromosome preparations from solid tissues of fishes. J. Fish. Res. Board Can. 34: 266–269.

Ogawa, M., M. Yagasaki and F. Yamazaki. 1973. The effect of prolactin on water influx in isolated gills of the goldfish, *Carassius auratus* L. Comp. Biochem. Physiol. 44A: 1177–1183.

Scheel, J.J. 1968. Rivulins of the Old World. T.F.H. Publications, Neptune City, N.J.

Scheel, J.J. 1972. Rivuline karyotypes and their evolution (Rivulinae, Cyprinodontidae, Pisces). Z. Zool. Syst. Evol. 10: 180–209.

Scheel, J.J. 1974. Rivuline studies. Taxonomic studies of rivuline cyprinodonts from tropical Atlantic Africa (Rivulinae, Cyprinodontidae, Atheriniformes, Pisces). Annales du Musée Royal de l'Afrique centrale 211. Tervuren, Belgique.

Scheel, J.J. 1990. Atlas of Killifishes of the Old World. T.F.H. Publications, Neptune City, N.J.

Völker, M., R. Sonnenberg, P. Ráb and H. Kullmann. 2006. Karyotype differentiation in *Chromaphyosemion* killifishes (Cyprinodontiformes, Nothobranchiidae). II: Cytogenetic and mitochondrial DNA analyses demonstrate karyotype differentiation and its evolutionary direction in *C. riggenbachi*. Cytogenet. Genome Res. 115: 70–83.

Völker, M., R. Sonnenberg, P. Ráb and H. Kullmann. 2007. Karyotype differentiation in *Chromaphyosemion* killifishes (Cyprinodontiformes, Nothobranchiidae). III: Extensive karyotypic variability associated with low mitochondrial haplotype differentiation in *Chromaphyosemion bivittatum*. Cytogenet. Genome Res. 116: 116–126.

7

Direct Chromosome Preparations from Embryos and Larvae

M. Völker[1] and *P. Ráb*[2,*]

[1]Institut für Evolutionsbiologie und Ökologie, Universität Bonn, 53121 Bonn, Germany.
e-mail: mv53121@googlemail.com

[2]Laboratory of Fish Genetics, Institute of Animal Physiology and Genetics, Academy of Science of the Czech Republic, 277 21 Liběchov, Czech Republic.
e-mail: rab@iapg.cas.cz

*Corresponding author

Principles

Direct chromosome preparation from fish embryos and larvae follows the basic protocol described by Kligerman and Bloom (1977). In short, whole eggs, embryos or larvae are directly incubated in aquarium water with colchicine, treated with a hypotonic solution, to allow the swelling of the cells. After fixation a cell suspension is prepared from each single embryo/egg and dropped onto clean slides.

The various methods using early developmental fish stages for direct chromosome preparations involve treatments of either whole eggs or embryos. For example, Phillips et al. (1985) and Ueda et al. (1991) removed embryos from the eggs before incubating them in culture medium containing colchicine, while Chourrout and Happe (1986) treated whole eggs with aqueous colchicine solution. According to our experience, the latter method normally yields entirely satisfactory results. Baksi and Means (1988) applied both mitostatic and hypotonic treatment to whole eggs, which were then fixed and used for the preparation of cell suspensions. While this protocol avoids dissecting the embryo out of the egg, it may hamper hypotonic swelling of the cells (Baksi and Means 1988); moreover in yolk-rich eggs, the yolk can interfere with certain staining and banding techniques, a problem that was already noted by Roberts (1967). Therefore, dissecting the embryo out of the egg and discarding the contents of the yolk sac prior to hypotonic treatment is probably advisable in most cases. Similarly, if larvae of the species under study have a large yolk sac, it should be removed before hypotonic treatment.

Regardless of whether embryos or larvae are used for chromosome preparation, two major alternatives exist with respect to the preparation of cell suspensions and slide preparation. Variant 1 (Kligerman and Bloom 1977) involves dissociating the cells in 50% aqueous acetic acid, dropping a small amount of the cell suspension onto heated slides and sucking it back into the pipette after some seconds, leaving a ring of cells on the slide. The metaphases are found at the periphery of this ring and are thus easy to locate. This technique allows preparation of several slides from a small amount of tissue. In variant 2, the tissues are dissociated in a mixture of acetic acid and ethanol or methanol in a proportion ranging from 1:1 to 1:3 (e.g., Ueda et al. 1991; Nomura et al. 2004), or in 100% acetic acid which is later mixed with the two- or threefold amount of methanol (e.g., Inokuchi et al. 1994). This cell suspension is dropped onto cold, clean slides. Compared to variant 1, this method may in some cases improve spreading.

Applications and Limitations

Chromosome preparation from embryos and larvae is a fast, simple and cheap method that permits the preparation of a high number of high-quality metaphases, especially if embryos are used. In addition to standard karyotype analysis (e.g., Völker et al. 2005), the method is particularly useful in studies on genotoxic effects on early life stages of fishes (e.g., Ueda et al. 1991) and in experiments on the artificial induction of polyploidy (e.g., Strüssmann et al. 1993; Felip et al. 1997; Nomura et al. 2004) and gynogenesis (Lin and Dabrowski 1996; Gomelsky et al. 1998). Chromosome preparations from embryos and larvae can routinely be used for the most widespread staining and banding techniques such as Giemsa staining, $AgNO_3$-staining, C-banding and fluorescence banding with CMA_3 and DAPI (Völker and Kullmann 2006; Fig. 1) as well as FISH (Fujiwara et al. 1998). Furthermore, the combination of chromosome preparation from embryos with protocols for BrdU incorporation yields high-quality replication banding (Inokuchi et al. 1994; Ueda and Naoi 1999).

Unlike most other protocols for direct chromosome preparation, the use of embryos and larvae does not require sacrificing adult specimens, which is particularly advantageous if specimens of the species under study are hard to get or must be used in additional experiments. Besides, the protocol given here permits chromosome preparation from single embryos and thus allows

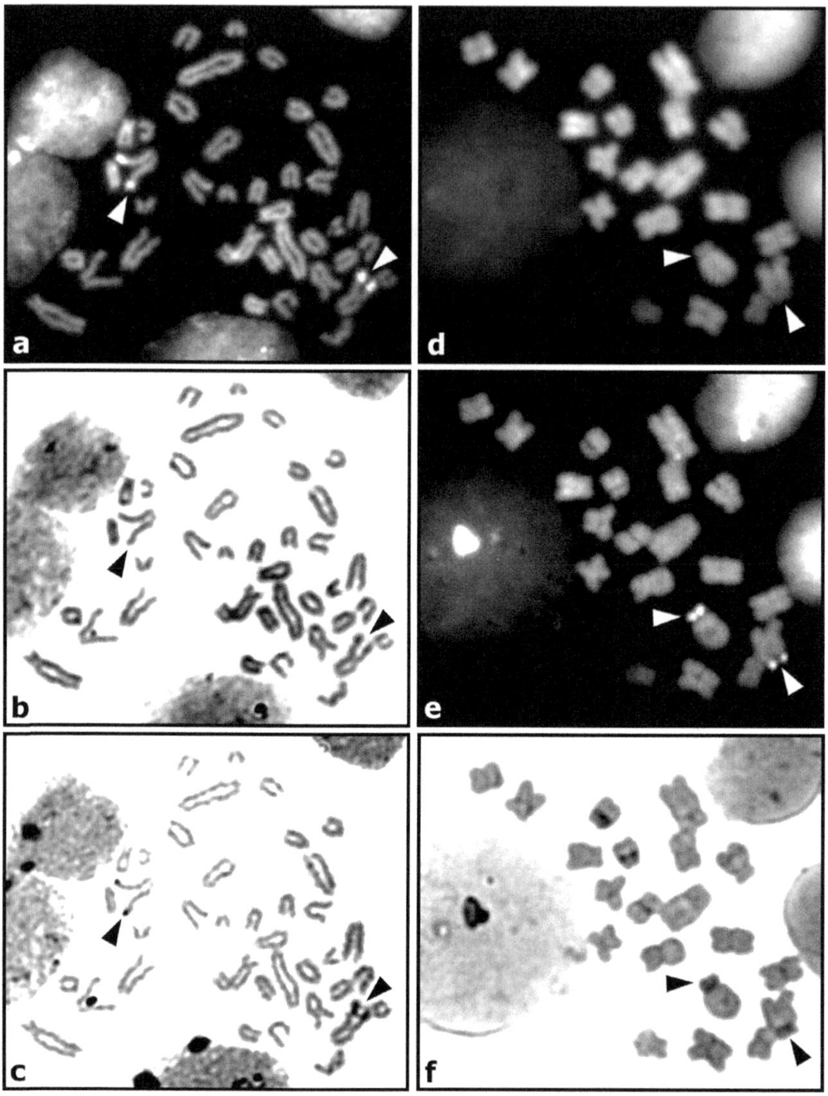

Figure 1. Metaphase chromosomes prepared from embryos of *Chromaphyosemion* killifishes. (a-c) Metaphase chromosomes of *C. poliaki* "Bowanda" (2n = 38) after (a) CMA_3-banding, (b) Giemsa staining and (c) $AgNO_3$-staining. (d-f) Metaphase chromosomes of *C. riggenbachi* "KV03/29" after (d) DAPI-banding, (e) CMA_3-banding and (f) C-banding. Arrowheads in (a-f) indicate Ag-NORs associated with GC-rich C-positive heterochromatin.

for the analysis of interindividual chromosomal variation. Disadvantages of the technique are that sex and viability of the specimens examined are unknown and obtaining embryos or larvae by breeding or artificial insemination is not always possible.

Protocol

A. Direct chromosome preparation from embryos

This protocol is a modification of the methods of Chourrout and Happe (1986), Ueda et al. (1991) and Völker and Kullmann (2006).

1. Incubate whole eggs (early eyed stage) in aquarium water with 0.05% colchicine for 4 hrs at optimum temperature for development in the species under study.
2. Dissect the embryo out of the egg in a watchglass containing a drop of the colchicine solution, using fine dissection needles. Open the yolk sac and discard the yolk.
3. Transfer the embryo with fine forceps to a 1.5 ml microtube filled with 0.8% trisodium citrate. Incubate for 30–50 min at the same temperature used for mitostatic treatment.
4. Using a Pasteur pipette, replace the trisodium citrate with freshly prepared, cold fixative (3:1 methanol : acetic acid) and incubate the microtube at 4°C for 20 min.
5. Using a Pasteur pipette, change the fixative at least twice and extend the last fixation step overnight.

Proceed by preparing a single cell suspension and dropping it on slides in one of the following ways:

Variant 1
(from Chourrout and Happe 1986 after Kligerman and Bloom 1977):

6. Remove the embryo from the microtube, blot it briefly to a piece of filter paper to remove excess of fixative and transfer it to a new 1.5 ml microtube containing a few drops of 50% aqueous acetic acid. Gently mince the embryo with a micropestle to produce a single-cell suspension.
7. Using a micropipette, place 45 µl of the cell suspension onto a clean, dry slide that has been heated to about 45°C on a slide warmer. After 20 s, suck back the suspension into the micropipette, leaving a ring of cells on the slide. Repeat this process twice per slide, creating three adjacent rings of cells.

Variant 2
(from Ueda et al. 1991):

8. Remove the embryo from the microtube, touch it to a piece of filter paper and transfer it to a new 1.5 ml microtube containing a few drops of 1:1 acetic acid: methanol. Gently mince the embryo with a micropestle.
9. Using a micropipette, drop 20 µl of the cell suspension onto a clean slide and leave it to dry.

B. Direct chromosome preparation from larvae

This protocol is a modification of the methods of Kligerman and Bloom (1977) and Baksi and Means (1988).

1. Place larvae into aquarium water with 0.05% colchicine for 4 hrs.
2. Open the yolk sac with fine dissection needles and remove its contents.
3. Proceed with steps 3–7 of the embryo preparation protocol.

Troubleshooting

The most common problems encountered with direct chromosome preparation from embryos and larvae are an insufficient amount of metaphases and improper spreading of chromosomes. The mitotic index strongly depends on the developmental stage. Studies in salmonids (Chourrout and Happe 1986) and killifishes (Völker and Kullmann 2006) suggest that the early eyed stage is most suitable for cytogenetic analysis, probably because of high mitotic activity due to tail bud formation and growth. The preparation of later embryonic stages or larvae yields substantially lower mitotic indexes.

Bad spreading of chromosomes may be overcome by a prolonged hypotonic treatment or by using a different hypotonic solution, for example 0.075 M KCl or distilled water. In our experiments with killifish embryos of the African genus *Chromaphyosemion*, hypotonic treatment with 0.075 M KCl resulted in faster spreading, more scattering and a higher contraction of chromosomes than hypotonic treatment with 0.8% trisodium citrate (Völker and Kullmann 2006). The most suitable hypotonic treatment, term of chemical composition of the solution and duration must be determined empirically according to the species under study.

If cell suspensions and slides are prepared according to variant 1, spreading can be improved by increasing the temperature of the slide on which the cell suspension is dropped, increasing the size of the rings of cells on the slide and decreasing the time that the suspension remains on the slide (Kligerman and Bloom 1977). Furthermore, good spreading of chromosomes generally requires a volume of hypotonic medium adequate to the amount of tissue/cells treated (Foresti et al. 1992). Thus, it may be necessary to use higher amounts of hypotonic solution according to the size of embryos and larvae under study.

A minor problem is under- or overtreatment with colchicine, which can be solved by appropriately adjusting the colchicine concentration and/or the duration of the mitostatic treatment.

Laboratory set up

Equipment

Major

Slide warmer, adjustable to 45°C; stereo-microscope; refrigerator.

Minor

Plastic Petri dishes (diameter 55 mm); fine dissection needles; watch-glasses; fine forceps; 1.5 ml microtubes; micropestle; filter paper; Pasteur pipettes; micropipette (10–100 μl) with tips; microscope slides; timer.

Biochemicals and reagents

Colchicine; $C_6H_5Na_3O_7$ or KCl; methanol (ice cold); glacial acetic acid; hypotonic solution (0.075M KCl).

Stock and working solutions

Hypotonic solution: trisodium citrate, 10x:

8% $C_6H_5Na_3O_7$ in distilled water.

Dilute 1:10 with distilled water before use.

References

Baksi, S.M. and J.C. Means. 1988. Preparation of chromosomes from early stages of fish for cytogenetic analysis. J. Fish Biol. 32: 321–325.

Chourrout, D. and A. Happe. 1986. Improved methods for chromosome preparation in rainbow trout, *Salmo gairdneri*. Aquaculture 52: 255–261.

Felip, A., S. Zanuy, M. Carrillo, G. Martinez, J. Ramos and F. Piferrer. 1997. Optimal conditions for the induction of triploidy in the sea bass (*Dicentrarchus labrax* L.). Aquaculture 152: 287–298.

Foresti, F., C. Ozouf-Costaz and P. Ráb. 1992. Volume dependent effect of hypotonic treatment on good spreading of fish chromosomes. Cybium 16: 209–212.

Fujiwara, A., S. Abe, E. Yamaha, F. Yamazaki and M.C. Yoshida. 1998. Chromosomal localization and heterochromatin association of ribosomal RNA gene loci and silver-stained nucleolar organizer regions in salmonid fishes. Chromosome Res. 6: 463–471.

Gomelsky, B., N.B. Cherfas, A. Gissis and G. Hulata. 1998. Induced diploid gynogenesis in white bass. Prog. Fish Cult. 60: 288–292.

Inokuchi, T., S. Abe, E. Yamaha, F. Yamazaki and M.C. Yoshida. 1994. BrdU replication banding studies on the chromosomes in early embryos of salmonid fishes. Hereditas 121: 255–265.

Kligerman, A.D. and S.E. Bloom. 1977. Rapid chromosome preparations from solid tissues of fishes. J. Fish Res. Board Can. 34: 266–269.

Lin, F. and K. Dabrowski. 1996. Effects of sperm irradiation and heat shock on induction of gynogenesis in muskellunge (*Esox masquinongy*). Can. J. Fish. Aquat. Sci. 53: 2067–2075.

Nomura, K., J. Nakajima, H. Ohta, H. Kagawa, H. Tanaka, T. Unuma, K. Yamauchi and K. Arai. 2004. Induction of triploidy by heat shock in the Japanese eel *Anguilla japonica*. Fisheries Sci. 70: 247–255.

Phillips, R.B., K.D. Zajicek and F.M. Utter. 1985. Q band chromosomal polymorphism in chinook salmon (*Oncorhynchus tshawytscha*). Copeia 1985: 273–278.

Roberts, E.L. 1967. Chromosome cytology of the Osteichthyes. The Progressive Fish-Culturist 29: 75–83.

Strüssmann, C.A., N.B. Choon, F. Takashima and T. Oshiro. 1993. Triploidy induction in an atherinid fish, the pejerrey (*Odontesthes bonariensis*). The Progressive Fish-Culturist 55: 83–89.

Ueda, T. and H. Naoi. 1999. BrdU-4Na-EDTA-Giemsa band karyotypes of 3 small freshwater fish, *Danio rerio*, *Oryzias latipes*, and *Rhodeus ocellatus*. Genome 42: 531–535.

Ueda, T., M. Hayashi, N. Koide, T. Sofuni and J. Kobayashi. 1991. Preliminary examination of the mutagenicity test using embryo cells of rose bitterling, *Rhodeus ocellatus ocellatus*. Chromosome Information Service 51: 12–14.

Völker, M. and H. Kullmann. 2006. Sequential chromosome banding from single acetic acid fixed embryos of *Chromaphyosemion* killifishes (Cyprinodontiformes, Nothobranchiidae). *Cybium* 30: 171–176.

Völker, M., P. Ráb and H. Kullmann. 2005. Karyotype differentiation in *Chromaphyosemion* killifishes (Cyprinodontiformes, Nothobranchiidae). I: Chromosome banding patterns in *C. alpha*, *C. kouamense* and *C. lugens*. Genetica 125: 33–41.

8

Establishment of Sturgeon Primary Cell Lines

F. Fontana

Department of Life Science and Biotechnology, University of Ferrara, Via L. Borsari 46, I-44100 Ferrara, Italy.
e-mail: francesco.fontana@unife.it

Principles

The most common techniques to obtain fish metaphase plates involve the cephalic part of the kidney as a source of haematopoietic tissue and peripheral blood leukocyte culture (Denton 1973; Blaxhall 1975; Gold et al. 1990; Klinkhardt 1991). These methods have two main limits: a) the quality of results may vary among different fish species (freshwater, marine, tropical or cold water); b) each individual may respond differently to treatments according to seasonal changes, age, sexual maturity and general physiological conditions. These difficulties may be mostly overcome by the establishment of primary cell lines from connective tissue explants.

The protocol starts with the sampling of tissue fragments from live fish kept in an aquarium or transferred to the laboratory in well-aired tanks. The tissue sampling does not require the sacrifice of the specimen (or even any serious discomfort or wounding of precious and strictly protected animals). From these small pieces of tissue we can easily set up cell culture lines from which good metaphase plates can be obtained at will.

Alternatively, if it is not possible to directly sample living animals in place, you can have them sampled elsewhere and the tissue fragments mailed to you. The new method for sterility sampling and long-distance mailing, allowed us to obtain very good results even when live tissue fragments of sturgeons were sampled and mailed from very long distances (from the United States to Italy, for example).

Applications and Limitations

Cell culture, although more time-consuming and difficult when compared to other techniques, has considerable advantages: i) once established, cultured cells can grow independently of physiological and environmental conditions, ii) in cultured cells it is possible to synchronize the cell cycle. Indeed, cell cultures provide good quality metaphases from which chromosomes can be precisely counted, morphologically characterized and treated for banding and *in situ* hybridization. Karyotype analyses are especially difficult in sturgeons, because the chromosome number is very high (some species are known to have up to 360 chromosomes), and there are numerous microchromosomes (see karyotype reconstructions and references at: http://www.unife.it/dipartimento/biologia-evoluzione/progetti/geneweb). The following protocol refers to a basic method employed in our laboratory, although it may not be suitable for all sturgeon species. Our experience suggests that details of the method should be adjusted not only when dealing with individuals of the same systematic group, but even with individuals of the same species. In the latter case, the physiological conditions of each animal play a relevant role (young individuals are preferred) (Fontana et al. 1995; Fontana et al. 1997).

Protocol

1. Biopsy (Fig. 1)

a) Chose a part to be cut from the dorsal fin (about 2 cm^2) and clean it very well with hydrogen peroxide (H_2O_2) solution for human use. Cut out the piece using sterile scissors, taking care to include a small portion of the underlying connective tissue (Fig. 1a).
b) After cutting, drop the piece of fin into a beaker with 70% ethanol, without touching the edge (sterile forceps) (Fig. 1b).
c) Grasping the fin piece with forceps, shake it gently in the ethanol solution to clean and sterilize the piece, for no more than 40 seconds (Fig. 1c).
d) Transfer under sterile conditions the fin piece into a capped sterile plastic tube, containing 0.9% NaCl, 100 I.U. penicillin, 0.1 mg/ml streptomycin and 0.1 mg/ml antifungal (such as fungizone). Be very careful not to contaminate the test tube mouth or cap (if necessary use a Bunsen flame) (Fig. 1d).
e) Close the tube very tightly and turn it slowly upside down 2–3 times to dilute the ethanol and soak the biopsy. From now on the test tube must be kept at 4–5°C (Fig. 1e).
f) The tubes containing the fin pieces must be sent by express mail (such as DHL, etc.) as quickly as possible, within a maximum of 24–48 hours in a common polystyrene box (about 30 x 30 x 20 cm) cooled by two refrigerating fluid containers (synthetic ice packets), perfectly sealed and

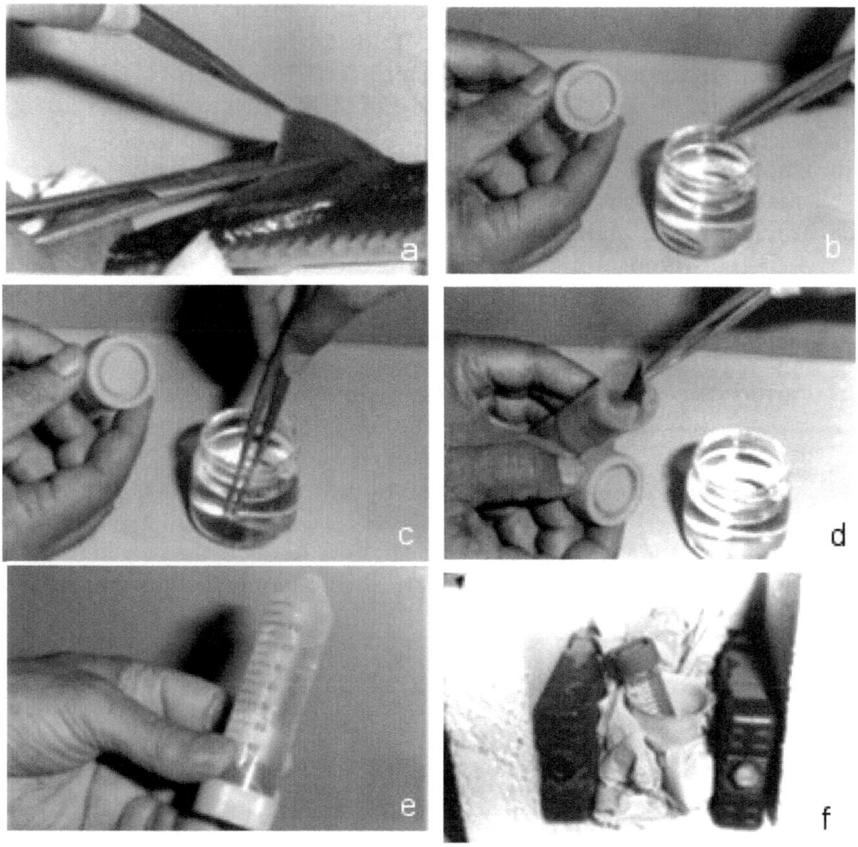

Figure 1. Biopsy from sturgeon dorsal fin for cell culture. (a) Dorsal fin cut with sterile scissors; (b) Fin piece dropped in beaker with 70% ethanol; (c) Fin piece sterilized in ethanol by gentle shaking; (d) Fin piece transferred to sterile test tube containing saline solution with antibiotics; (e) Dilution of ethanol by slow upturning of the test tube; (f) Correct packaging of test tube(s) in refrigerated polystyrene box.

previously kept at –20°C. The test tubes should be individually wrapped by insulating material (even common paper), to avoid direct contact with the synthetic ice packets. The polystyrene box should be large enough to contain the samples and the refrigerating packets placed on each inner side of the box, as shown in Fig. 1f.

2. Cell cultures

a) The primary cultures start from connective tissue explants from the dorsal fin, or from subcutaneous or muscular tissue. Before removing the tissue piece the outside surface must be swabbed with 70% ethanol.

b) Possibly under a laminar flow box (if available) cut with sterilized scissors tissue pieces of about 1 mm in size and sterilize their surface by soaking them in two 5-minute changes of sterilized CMF-PBS (pH 7.2) containing 200 IU/ml of penicillin, 200 µg/ml of streptomycin and 6.25 µg/ml of fungizone.
c) Place the explants into an empty sterile Petri dish and mince them into small fragments with sterile scalpel blades.
d) Transfer the small fragments into sterile plastic flasks (25 cm^2).
e) Adherence is encouraged by allowing the fragments to dry on the bottom for 10 minutes. Gently cover the fragments with 1–2 ml of growth medium.
f) Place the flasks in a 25°C incubator with 5% CO_2 atmosphere in a horizontal position.
g) Periodically examine the flasks for any cellular growth around the explants (inverted microscope, magnification from 20 to 40 x) and renew the medium when pH indicator changes colour.
h) After 2–4 weeks a layer of cells should appear around the explants (Fig. 2).
i) As soon as the layers of cells appear, add 10 mg/ml of basic Fibroblast Growth Factor (bFGF, Roche Diagnostics, Mannheim, Germany).

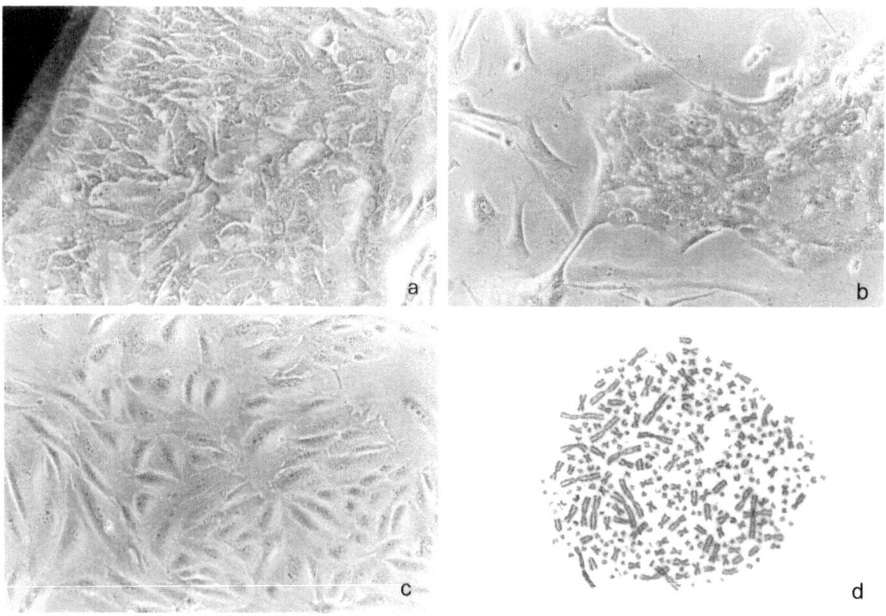

Figure 2. Phase contrast micrographs of cells grown from skin explants (a, b, c) and a metaphase plate (d) of *Acipenser baerii*. (a) Heterogenous cell population grown from skin explant (a fragment is visible in the upper left corner); (b) Selection of fibroblast-like cells (on the left) from a receding heterogeneous cell population (on the right); (c) Final fibroblast monolayer; (d) Metaphase plate of *A. baerii* (2n = 240). (a, b, c, X 200; d, X 1800).

j) When monolayer growth reaches about 70% confluence, discard the liquid medium from the flask, wash the cells for 10 min in CMF-PBS, gently discard the medium and add enough trypsin-EDTA working solution to cover the bottom of the flask.
k) Incubate at room temperature for 1–2 minutes, checking under an inverted microscope for when cells are beginning to round up and detach.
l) Re-suspend the detached cells in 5 ml of growth medium enriched with 10% foetal calf serum (FCS) (the serum is necessary to neutralize the trypsin) and centrifuge (10 min at 500 rpm). If there are any explants still attached to the flask, then they must be maintained in culture by adding new medium (more cell layers may be produced).
m) Centrifuge the detached cell suspension (10 min at 500 rpm), resuspend the pellet in growth medium and transfer into new flasks, adding a suitable amount (5 ml or more) of growth medium. Generally from the first flask about 4–6 new flasks are obtained. Subcultures are obtained dividing the cells collected in one flask in two new ones, by the above described technique.
n) After two or three subcultures, as soon as the first groups of fibroblast-type cells appear, a brief wash in distilled water is performed, after which new growth medium is added.
o) The cell cultures, all made of fibroblasts, are allowed to grow for a further 2–3 weeks, regularly changing the culture medium every 3–4 days, until cells reach confluence for two-thirds of the flat space available.
p) In order to reduce the number of subcultures and thus avoid the risk of obtaining heteroploid karyotypes, the cells should be collected and frozen after about eight subcultures.

3. Cultured cell storage

Collect cells at the maximum growth (identified by inverted microscope observations) and after centrifugation resuspend them in FCS-preservation medium (3:1) in order to obtain a concentration of about 10^6–10^7 cells/ml. The counting of cells in the cell suspension can be performed by an hemocytometer. Place the cell suspension in freezing vials (about 2 ml each). Cool the vials gradually: store them in a refrigerator for an hour and then in freezing steel cans in liquid nitrogen.

4. Cultured cell recovery

a) Remove the vials from frozen storage and thaw them for 2–3 minutes in a water bath at 3°C, slowly and continuously whirling each vial to mix cells.

b) Open the vial in sterile conditions, sample about 0.5 ml of cell suspension with a sterile pipette and place it into a culture flask with culture medium enriched with 20% FCS. Maintain flasks in a 25°C incubator with 5% CO_2 atmosphere.

5. Chromosome preparations

a) When the cultures reach 70–80% confluence, add colchicine (final concentration, 0.5 µg/ml) for six hours.
b) Add 5 ml trypsin-EDTA working solution to the flask. Detach cells by tilting the flask sideways.
c) Examine the flask under an inverted microscope. The cells should detach from the bottom (if not, incubate for a few more minutes).
d) Add 5 ml CMF-PBS to stop trypsin treatment and transfer the suspension into a centrifuge tube.
e) Centrifuge the tube at 800 rpm for 10 minutes. Discard the supernatant, leaving as little medium as possible over the pellet.
f) Resuspend the pellet in 5 ml of in hypotonic phosphate buffer solution for 20–30 minutes at room temperature.
g) In our case (but maybe not in all cases) we found it useful to perform a pre-fixation by adding 5–6 drops of ice-cold fixative to the 5 ml of hypotonic phosphate buffer solution containing the cells. Gently mix.
h) Centrifuge at 800 rpm for 10 minutes. Discard the supernatant, slowly add 5 ml of cold fixative to the pellet and gently mix.
i) Centrifuge again under the same conditions, discard the supernatant and resuspend the cell pellet in 5-ml fresh ice-cold fixative. Repeat this step a third time.
j) After the final centrifugation, resuspend the cells in a small volume of fixative (0.5 to 1 ml, depending on the pellet size).
k) Allow 3 to 4 drops of the suspension to fall from a distance of about 50 cm onto a cold, wet and clean glass slide, and allow to dry at room temperature.

Troubleshooting

One of the initial problems we faced was the failure to obtain good metaphase plates from the first generation of cells developing from explants. The phase contrast micrography in Fig. 2a shows, in the upper left corner, an explant of *Acipenser baerii* skin: after about 20 days of culture a heterogeneous cell population is still clearly present. Thus it was very difficult to use these first-generation cells to obtain good metaphase plates, since each cell subpopulation was in a different stage of cell cycle. The complexity of the cell population had to be eliminated by promoting proliferation of fibroblast-like strains, more resistant and easier to manipulate. For this we employed three different techniques. The simplest one was to continue to grow the cells until original populations were naturally replaced with fibroblast-like ones: this unfortunately required a long time, increasing the risk of aneuploidy. However, we noticed in the initial trials that, by further diluting the culture medium, many cell types weaker than fibroblasts underwent lysis: therefore, as soon as the first groups of fibroblast-like cells appeared, short washes with distilled water instead of PBS could easily eliminate the other cells by lysis. More recently we found that the addition of a specific growth factor for fibroblasts (bFGF) to the culture medium was able to speed up fibroblast growth, so they could overcome all other cell populations. The right part of Fig. 2b shows a group of slowly disappearing heterogeneous cells, while the left shows a group of fibroblast-like ones, actively proliferating. After about 6–8 subcultures, we obtained a homogeneous layer of fibroblasts (Fig. 2c): meanwhile the fibroblast doubling time gradually reduced to 24–48 hr. The presence of the homogenous fibroblast layer makes it possible to safely block cell growth and obtain good metaphase plates (Fig. 2d).

All sturgeon cell lines obtained by the above technique were maintained in our laboratory for more than 15 subcultures. Although cell lines are usually accompanied by genome alterations, the modal chromosome number of our cell lines did not change, especially because we decided to freeze some of them in liquid nitrogen after about only 2–3 subcultures.

Although in our laboratory the interest to establish euploid sturgeon cell lines stemmed from the need to obtain metaphase plates suitable for cytogenetic studies, these cell lines could certainly be useful for other research purposes, including icthyopathology studies.

Laboratory set up

Equipment

Major

Aquarium; laminar flow box; centrifuge; autoclave; hemocytometer; standard CO_2 incubator; inverted microscope; household 4°C refrigerator; aspiration pump.

Minor

Sterile dissecting tools; angle necked sterile cell culture flasks, with 25 cm^2 flat bottom surface; sterile pipettes (1 ml, 10 ml); cell culture tubes 16 x 125 mm; cryovials (2 ml, 4 ml); 10 cm Petri dishes, liquid nitrogen storage system.

Biochemicals and reagents

Dulbecco/Vogt modified Eagle's minimal essential medium (DMEM) with high glucose; 4-(2-hydroxyethyl) -1-piperazine ethanesulfonic acid (HEPES) buffer; sodium bicarbonate; foetal calf serum (FCS); 100 I.U./ml penicillin and 100 µg/ml streptomycin; L-glutamine; sodium chloride; potassium chloride; sodium phosphate; potassium phosphate; trypsin; EDTA; dimethylsulfoxide (DMSO); D-glucose; absolute methanol; water-free acetic acid; distilled water; Carnoy's fixative.

Stock and working solutions

10 X Calcium- and Magnesium-Free phosphate buffered saline (CMF-PBS): (8% NaCl, 0.2% KCl, 1.44% $Na_2HPO_{4,}$ 0.24% KH_2PO_4).

Growth medium: DMEM High glucose, 20 mM HEPES, 10 mM $NaHCO_3$, 15% FCS, 100 I.U/ml penicillin and 100 µg/ml streptomycin, 0.1 mg/ml antifungal (fungizone), 20 mM L-glutamine, pH 7.2.

EDTA stock solution, 2% (pH adjusted at 7.4 with NaOH 2N).

Trypsin stock solution, 2.5% in PBS (dissolve 2.5 g trypsin in 100 ml of ice-cold physiological solution).

Trypsin–EDTA solution: to 2 ml trypsin stock solution add 0.5 ml of EDTA stock solution. Add physiological solution to 10 ml. Sterilize by filtration.

Preservation medium: Growth medium 48%, DMSO 40%, D-glucose 12% w/v.

Physiological solution: 0.9% NaCl.

Sorensen's phosphate buffer (0.15 M, pH 6.8): Na_2HPO_4, 0.702 %; $NaH_2PO_4\ 2H_2O$, 0.836 %.

Hypotonic phosphate buffer solution: Sorensen's phosphate buffer 20%.

Acknowledgements

The author wish to thank Dr. Massimo Lanfredi and Dr. Milvia Chicca, from the Department of Life Sciences and Biotechnology, University of Ferrara, respectively for valuable technical suggestions and for critically reviewing the text. This work was supported by Italian Ministry of University and Research (MIUR) grant n. DM219-22-12-05, Nationally Relevant Research Projects (PRIN), 2005.

References

Blaxhall, P.C. 1975. Fish chromosome techniques–a review of selected literature. J. Fish Biol. 7: 315–320.

Denton, E.T. 1973. Fish Chromosome Methodology. Charles Thomas Publisher, Springfield.

Fontana, F., M. Lanfredi, R. Rossi, P. Bronzi and G. Arlati. 1995. Established cell lines from three sturgeon species. Sturgeon Quarterly 3(4): 6–7.

Fontana, F., R. Rossi, M. Lanfredi, G. Arlati and P. Bronzi. 1997. Cytogenetic characterization of cell lines from three sturgeon species. Caryologia 50: 91–95.

Gold, J.R., Y.C. Li, N.S. Shipley and P.K. Powers. 1990. Improved methods for working with fish chromosomes with a review of metaphase chromosome banding. J. Fish Biol. 37: 563–575.

Klinkhardt, M.B. 1991. A brief comparison of methods for preparing fish chromosomes: an overview. Cytobios 67: 193–208.

9

Teleost Fish Lymphocyte Culture

S. Salvadori,[a,*] *E. Coluccia*[b] and *A.M. Deiana*[c]

Dipartimento di Scienze della Vita e dell'Ambiente, Università di Cagliari, Via T. Fiorelli 1, 09126 Cagliari, Italy.
[a]e-mail: salvador@unica.it
[b]e-mail: coluccia@unica.it
[c]e-mail: adeiana@unica.it
*Corresponding author

Principles

Lymphocyte culture is a short term cell culture technique used to obtain a large number of metaphases necessary for cytogenetic investigations. This technique was first used for culturing human blood in the 1960's (Boll and Fuchs 1961; Woodliff 1964). In fishes it was first successfully used in *Cyprinus carpio* in 1967 by Labat et al. The principle is based on the incubation of either the whole blood or leukocytes in a suitable medium in the presence of components stimulating cell division (mitogens). Mitogens activate the immunocompetent cells thus stimulating cell growth and division (Pfeiffer 1974). Phytohemagglutinin (PHA), extracted from the red kidney bean *Phaseolus vulgaris*, first used for separating leukocytes from whole blood by agglutination of the erythrocytes, has been demonstrated to induce leukocyte division (Nowell 1960) and it is by far the most used mitogen.

Applications and Limitations

The technique can provide cell suspensions both from freshwater and from marine teleost fish species, producing medium to high mitotic index (Denton 1973); it is the most used cell culture method in cytogenetics (Rooney and Czepulkowski 1986). The chromosome preparations contain few cell debris and are thus suitable for the applications of subsequent cytogenetic protocols. As an important advantage, the fishes are not sacrificed and may be kept in aquaria; this is useful in case of matrices in fish farms, and when dealing with protected or rare species. Moreover several blood cultures can be obtained from the same specimen.

The most important limitation of the method is the size of the fishes since it can be impossible to obtain the minimum volume of blood for the culture from small species or small specimens. Another limitation is the inter-individual variability in the mitotic response of cultured cells, as a general feature in fishes and other vertebrates (Rooney and Czepulkowski 1986; Fujiwara et al. 2001). The best results are obtained from cells cultured within 24 hours after sampling; however, lymphocytes can be preserved at 4°C up to 96 hours before culturing (Fujiwara et al. 2001). Many protocols on fish lymphocyte culture are available, and different combinations of parameters have to be tested for each species.

Protocol

All procedures connected with the blood cell culture require rigorous sterile conditions.

1. Culture medium preparation

Two principal protocols are proposed:

A) According to Fujiwara et al. (2001):
   199 Medium
   10% Foetal bovine serum
   60 μg/ml kanamycin sulfate
   1x antibiotic-antimycotic solution (100 U/ml penicillin, 100 μg/ml streptomycin, 250 ng/ml of amphotericin B)
   25 μM mercaptoethanol
   18 μg/ml phytohemagglutinin W form (PHA-W) and 100 μg/ml lipopolysaccharide (LPS).

B) The protocol provided here has been successfully tested by the authors in Anguilliformes (Cau et al. 1988; 1992; Deiana et al. 1990; Salvadori et al. 1994; 1995; 2003; 2009) and Perciformes (Deiana et al. 2000; Coluccia et al. 2010):
   RPMI 1640 L-Glutamine free
   20% Foetal Bovine Serum (GIBCO, origin: South America)
   1% Penicillin Streptomycin (10,000 U/ml: 10,000 μg/ml)
   1% Phytohemagglutinin M Form
   1% L-Glutamine 200 mM (L-glutamine easily degenerates; therefore it is better to add it to the culture medium just before use).

2. Collecting blood

- anesthetize the specimens in water containing 3-aminobenzoic acid ethyl ester (0.2 g/l).
- collect blood by caudal vein puncture using a heparinized syringe after carefully cleaning the skin with 70% ethanol.

3. Blood cell culture

Lymphocytes can be cultured without previous separation of erythrocytes as whole blood culture (A) or after separation (B). The advantage of protocol B is that preparations do not contain the erythrocyte nuclei, which are present in red-cells of non-mammals vertebrates (Fig. 1); but, in general, the mitotic index obtained is lower.

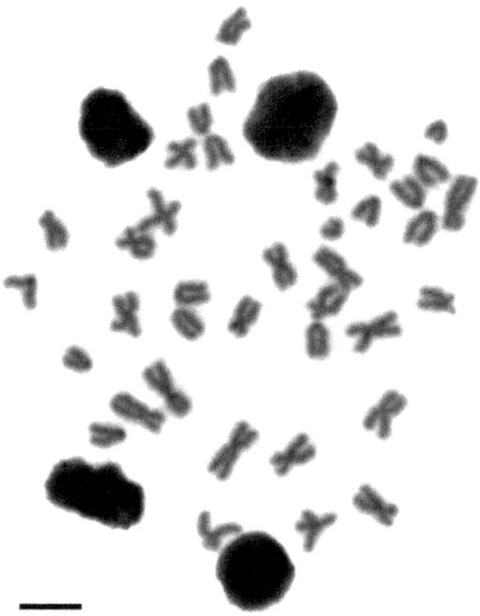

Figure 1. Metaphase of *Anguilla anguilla*, 2n = 38, after whole blood culture and Wright's staining. The scale bar indicates 5 μm.

A. Whole blood culture

The simplest method is the incubation of a small amount of whole blood (0.20–0.25 ml corresponding to about 4 to 5 drops of a 2.5 ml syringe) in 5 or 10 ml of complete culture medium in sterile conditions, under biological safety incubator.

B. Lymphocyte culture after separation of erythrocytes

To separate the leukocytes from erythrocytes the following methods are available:

i) Add the whole blood contained in a heparinized syringe to 0.01% phytohemagglutinin M form, and keep for 2 hrs at room temperature to allow spontaneous sedimentation of erythrocytes (Moorhead et al. 1960). As an alternative, transfer the whole blood + PHA in a centrifuge test tube and centrifuge at 500 rpm for 15 min. The agglutinated erythrocytes sediment at the bottom of the tube and leukocytes form a separate layer over erythrocytes. Collect leukocytes with a sterile Pasteur pipette and incubate in complete culture medium. A small percentage of erythrocytes is unavoidable and it is not harmful.
ii) According to Sánchez et al. (1990): dilute 3 ml of blood in Hank's balanced salt solution (HBSS) 1:1 and centrifuge twice at 40 g for 5 min at 4°C. Lymphocyte rich plasma is centrifuged at 100 g for 10 min at 4°C. After serum elimination, resuspend cells and wash twice in HBSS at 100 g for 10 min at 8°C.
iii) According to Fujiwara et al. (2001): mix 0.8 ml of heparinized blood with 5 ml of cold HBSS or culture medium in a 15 ml sterile plastic tube, placed on ice for 5 min and centrifuge at 1200 rpm for 5 min at room temperature. After centrifugation collect the lymphocyte rich plasma, centrifuge, and re-suspend the pellet in 5 ml of complete culture medium.

Separation in a gradient of macromolecular substances (Percoll, Ficoll, Dextran) may also be used but the results are not satisfactory in terms of mitotic index obtained.

C. Cell culture and colcemid treatment

Either the whole blood or lymphocytes are cultured in 10 ml or 5 ml of complete culture medium at 19°C to 28°C for 3 to 6 days.

1.5 to 5 hrs before collecting cultivated cells, colcemid (a semisynthetic derivative of colchicine) is added. Different combinations of temperature, time of culture and colcemid incubation can be used as follows, according to the physiology (temperature tolerance, metabolism) of the species under study:

i) 28°C, 72 hrs of culture, colcemid to a final concentration of 0.12 μg/ml, for 1.5 hr (Salvadori et al. 2003).
ii) 19°C, 5 days of culture, colcemid to a final concentration of 0.0125 μg/ml, for 5 hrs (Sánchez et al. 1990).
iii) 20°C, six days of culture, colcemid to a final concentration of 0.5 μg/ml, for 2.5 hrs (Fujiwara et al. 2001).

4. Harvesting cultured cells

At the end of the culture period, transfer the cultivated cell suspension to a conical tube.

- Centrifuge at 1200 rpm for 6 min.
- Discard the supernatant.
- Resuspend the pellet in 10 ml of hypotonic 0.075 M KCl solution, pre-warmed at 32°C in a water bath. Gently mix using a Pasteur pipette, by using the stem pipette only, to avoid the loss of cells in the pipette bulb, and incubate in a water bath at 32°C for 30 min.
- Centrifuge at 1000 rpm for 6 min and discard the supernatant.
- Fix cells with the alternative following methods:
  i) Add 10 ml of cold, freshly prepared Carnoy fixative (3 parts methanol or absolute ethanol: 1 part glacial acetic acid); the fixative must be added drop by drop at first, and carefully mixed with a Pasteur pipette to avoid lumps. Centrifuge at 1200 rpm for 6 min. These steps should be repeated once or twice.
  ii) The method of Terzoli et al. (1988) provides clean cell suspensions:
  iii) Resuspend the cells in 10 ml of Ibraimov mixture and centrifuge at 1000 rpm for 6 min.
  iv) Discard supernatant.
  v) Resuspend cells in methanol and centrifuge at 1000 rpm for 6 min.
  vi) Discard supernatant and fix cells in 3:1 methanol or absolute ethanol: acetic acid and centrifuge at 1200 rpm for 6 min.

5. Slide preparations

After the final fixation the cell suspension can be stored at –20°C or spread onto cleaned slides as follows:

- After centrifugation resuspend the pellet in a small volume of fixative, drop at arm length with a Pasteur pipette three to four drops of the suspension onto cleaned slides kept on absolute ethanol and dried just before use.
- Check the spreading density under phase contrast microscope and, if necessary, dilute or concentrate the cell suspension. If a cytoplasm halo is present around metaphase plates, the suspension could be dropped on wet slides. In this case, the metaphase suspension is dropped from a height of about 10 to 20 cm onto slides covered by a thin layer of Carnoy fixative and immediately laid down for few seconds on a horizontal support placed over boiling water to allow the fixative to evaporate; the excess of fixative is absorbed with tissue paper. Pay attention to leave the slides heating for just a few seconds.

6. Staining

Slides can be stained in 3:1 Sorensen phosphate buffer (pH 6.8): Wright solution for 6 min or in 5% Giemsa in the same buffer, for 10 min.

Troubleshooting

Cell division stimulation is one of the critical steps in culturing fish lymphocytes and it is important to test different mitogens. The most used are: phytohemagglutinin M form (PHA), Lipopolysaccharide (LPS), Pokeweed Mitogen (PWM) extract from pokeweed *Phytolacca americana* and Concanavalin A (ConA). Fujiwara et al. (2001) analyzed various combinations of parameters to optimize culture conditions for fish lymphocytes. They found a remarkable increasing in mitotic response by using a combination of PHA W + LPS in rainbow trout and other freshwater and marine species.

An additional possible problem is the foetal calf serum because of the variability of different batches; so that it can be necessary to test various batches to find the one which works.

Laboratory set up

Equipment

Major

Analytical balance; pHmeter; centrifuge; biological safety cabinet; water bath; refrigerated incubator; microscope; –20°C freezer; magnetic stirrer.

Minor

Sterile culture flasks: 12.5 to 25 cm^2 ml flasks; conical 15 ml centrifuge tubes; Pasteur pipettes; graduated pipettes; 2.5 to 5 ml syringes; microscope slides; microscope coverslips; sterile graduated pipettes; graduated measuring cylinders; beakers; timer.

Biochemicals and Reagents

3-Aminobenzoic acid ethyl ester methanesulfonate SIGMA; Heparin 50,000 U; Colcemid; Giemsa, 5%.

Stock and working solutions

IBRAIMOV MIXTURE: 92 ml distilled water, 3 ml methanol, 5 ml acetic acid

WRIGHT'S SOLUTION

Stock solution: Wright's powder is dissolved in methanol (0.25 g in 100 ml) in a magnetic stirrer in the dark; the solution is stabilized for 48 hrs at 37°C. Working solution: Wright's solution is diluted 1:3 in phosphate buffer described in the annex at the end of the book at pH 6.8.

References

Boll, I. and G. Fuchs. 1961. Vereinfachtes verfharen zur kurzfristigen kultivierung von menschlichem knochenmark *in vitro*. Blut. 7: 257.

Cau, A., S. Salvadori, A.M. Deiana, J.L. Bella and R. Mezzanotte. 1988. The characterization of *Muraena helena* mitotic chromosomes: karyotype, C-banding, nucleolar organizer regions and *in situ* digestion with restriction endonucleases. Cytogenet. Cell Genet. 47: 223–226.

Cau, A., E. Coluccia, A.M. Deiana, G. Pichiri, R. Rossino, S. Salvadori and R. Mezzanotte. 1992. Chromosomes and DNA of *Anguilla anguilla* L.: a study with restriction endonucleases. Genome 35: 838–843.

Coluccia, E., A.M. Deiana, A. Cau, R. Cannas, J. Tagliavini, A. Libertini and S. Salvadori. 2010. Karyotype analysis of the pumpkinseed *Lepomis gibbosus* (Actinopterygii, Centrarchidae) by chromosomal banding and *in situ* hybridization. J. Appl. Ichthyol. 26: 445–448.

Deiana, A.M., S. Salvadori and A. Cau. 1990. The characterization of mitotic chromosomes of *Gymnothorax unicolor* (Delaroche, 1809) by C-banding and NOR staining (Osteichthyes, Anguilliformes). Genetica 81: 17–20.

Deiana, A.M., A. Cau, S. Salvadori, E. Coluccia, R. Cannas, A. Milia and J. Tagliavini. 2000. Major and 5S ribosomal sequences of the largemouth bass *Micropterus salmoides* (Perciformes, Centrarchidae) are localized in GC-rich regions of the genome. Chromos. Res. 8: 213–218.

Denton, T.E. 1973. Handling of fish chromosomes. pp. 24–28. *In*: Denton, T.E. (ed.). Fish Chromosome Methodology. Charles C. Thomas Publisher, Illinois, USA.

Fujiwara, A., C. Nishida-Umehara, T. Sakamoto, N. Okamoto, I. Nakajama and S. Abe. 2001. Improved fish lymphocyte culture for chromosome preparation. Genetica 111: 77–89.

Labat, R., G. Larrouy and L. Malaspina. 1967. Technique de culture des leucocytes de *Cyprinus carpio* L. C.R. Acad. Sc. Paris 264: 2473.

Moorhead, P.S., P.C. Nowell, W.J. Mellman, D.M. Battips and D.A. Hungerford. 1960. Chromosome preparations of leukocytes cultured from human peripheral blood. Exp. Cell Res. 20: 613–616.

Nowell, P.C. 1960. Phytohemagglutinin: an initiator of mitosis in cultures of normal human leukocytes. Cancer Res. 20: 462.

Pfeiffer, R.A. 1974. Cell culture from blood and bone marrow. pp. 1–28. *In*: Schwarzacher, H.G. and U. Wolf (eds.). Methods in Human Cytogenetics. Springer-Verlag, Berlin, Heidelberg, New York.

Rooney, D.E. and B.H. Czepulkowski. 1986. Human Cytogenetics: a Practical Approach. IRL Press, Oxford, England.

Salvadori, S., A. Cau, E. Coluccia, A. Milia and A.M. Deiana. 1994. Karyotype, C- and G-banding and nucleolar organizer regions of *Conger conger* L. (Osteichthyes, Anguilliformes). Boll. Zool. 61: 59–63.

Salvadori, S., A.M. Deiana, E. Coluccia, G. Floridia, E. Rossi and O. Zuffardi. 1995. Colocalization of (TTAGGG)n telomeric sequences and ribosomal genes in Atlantic eels. Chrom. Res. 3: 54–58.

Salvadori, S., E. Coluccia, R. Cannas, A. Cau and A.M. Deiana. 2003. Replication banding in Mediterranean moray eels: chromosomal characterization and comparison. Genetica 119(3): 253–258.

Salvadori, S., E. Coluccia, R. Cannas, A. Cau and A.M. Deiana. 2009. A ZZ-ZW sex chromosome system in the finless eel *Dalophis imberbis* (Anguilliformes, Ophichtidae). Genetica 135: 283–288.

Sánchez, L., P. Martínez, A. Viñas and C. Bouza. 1990. Analysis of the structure and variability of nucleolar organizer regions of *Salmo trutta* by C-, Ag-, and restriction endonuclease banding. Cytogenet. Cell Genet. 54: 6–9.

Terzoli, G., A. Molinari, S. Stiovi, M. Buscaglia, G. Zuliani and G.E. Simoni. 1988. Second trimester foetal blood karyotyping in 85 malformed fetuses revealed by ultrasound. Proceedings of 21th Symposium on Cytogenetics, Prague, Vol. XXI.

Woodliff, H.J. 1964. Blood and Bone Marrow Cell Culture. Eyre & Spottiswoode, London.

10

Rapid Fibroblast Culture for Teleost Fish Karyotyping

M. Rábová,[1] *R. Monteiro,*[2,b] *M.J. Collares-Pereira*[2,c] and *P. Ráb*[1,a,*]

[1]Laboratory of Fish Genetics, Institute of Animal Physiology and Genetics, Academy of Sciences of the Czech Republic, 277 21 Liběchov, Czech Republic.
[a]e-mail: rab@iapg.cas.cz

[2]University of Lisbon, Faculty of Sciences, Centre for Ecology, Evolution and Environmental Changes, Campo Grande, C2, 1749-016 Lisboa, Portugal.
[b]e-mail: ritammonteiro@gmail.com
[c]e-mails: mjpereira@fc.ul.pt
*Corresponding author

Principles

The protocol for fibroblast culture from fish fin presented in this chapter is based on the method by Ojima (1978; 1982) later modified by Amemiya et al. (1984). Briefly, pieces of fins are sterilized, and placed as aseptically as possible into culture flasks using a suitable medium with adjusted pH, antibiotic/antimycotic agents and fetal calf serum. The cell culture is maintained at the temperature appropriate to the species under study, monitored daily for checking the explant outgrowth and possible contamination. When a monolayer of fibroblast cells covers the flask bottom, attaining around 70–80% of confluence, the cells are treated with colchicine and detached from the bottom surface by trypsin application. The cells are harvested and then hypotonized and fixed according to conventional procedure for chromosome preparations. The first protocols were based on the use of bicarbonate-buffered media (Eagel's Medium 199), which require frequent changes of the medium or culture in a CO_2-enriched atmosphere. Our modification applies L-15 medium, which is self-buffered by its amino acid composition and does not imply frequent changes, thus significantly reducing the manipulations and the consequent risk of contamination.

Applications and Limitations

Using cultured fibroblasts as a source of chromosomes has the major advantage that the fishes do not have to be sacrificed. This is especially important if the specimens are rare and/or protected, difficult and/or expensive to obtain, such as living in remote areas, or exhibited in public aquaria. Moreover the specimens remain available for possible additional examinations, such as crossing experiments, blood samples for leukocyte culture, biopsies for fibroblasts culture (see Chapter 8).

Although no experimental data are presently available, the fin samples can be possibly obtained in the field and stored in complete medium with increased dosage of antibiotic/antimycotic agents for some time before culturing. In addition to the above advantages, as in other cell cultures, the present protocol allows researchers to obtain a large number of prometaphases and metaphases, with high quality of chromosomes, thus making it possible to perform detailed cytogenetic analyses, including sequential banding and localization of DNA markers through molecular cytogenetic approaches.

The harvested cells can be stored in freezing medium in liquid nitrogen for possible recovery and setting up a new culture.

Cells from cultures can be also used for DNA extraction and/or flow cytometry.

On the other hand, cell cultures are much more expensive, time-consuming and require more laboratory facilities and equipment as compared to any other type of direct chromosome preparation. For routine applications, laboratory skills and experience are necessary to avoid contamination when setting up the cultures and during manipulations. As one of the major limitation, sub-cultures (i.e., passages to new flasks to allow new generations of cells) may result in chromosomal aberrations, including morphological modifications, changes in ploidy levels and formation of neo-chromosomes. Unless special cytogenetic studies of cell lines/cultures are planned, using the first or at most the second culture passage is therefore recommended.

Chromosomes from fin fibroblast cultures are generally more sensitive to hypotonic treatment and to denaturation than those obtained from direct preparations.

The success of fin fibroblasts culture depends on the species/group under study. The approach described here has been developed and successfully applied by our laboratory in cyprinids. Culture time and temperature provided in this protocol are specific for this fish group and may be modified and adapted according to the physiological requirements for other teleost fish groups. Both Ojima (1978; 1982) and Amemiya et al. (1984) also demonstrated successful results of fin fibroblasts cultures in cyprinid fishes. Although our experience with other fish groups is limited, some results of lower quality were obtained in cobitid loaches, umbrids, esocids and some catfish species. Therefore modification, combination and testing of different conditions given in the present protocol are required for routine and larger-scale application in other fish species.

Protocol

Modified from Ojima (1978; 1982), Amemiya et al. (1984) and Fontana et al. (1995).

1. Preparing the working space

a) Turn on UV light in the laminar flow box at least two hours prior to the experiment.
b) Turn off UV light, open the cabinet and wipe down with 70% ethanol.
c) Let defreeze and warm up the various components of the culture medium in the flow box, ensure that UV light was turned off (L-15, serum, antibiotic/antimycotic solution, L-glutamine). The serum can be thawed in a 37°C bath.

2. Preparing the complete medium

a) Wash hands carefully with an antiseptic soap or solution. Wipe the working area in flow-box with paper tissue wetted in 70% ethanol.
b) Check whether all necessary items and sterilized tools are in flow-box.
c) Mix 80 ml medium L-15 with 20 ml of serum in a graduated cylinder. Add 0.5 ml of antibiotic/antimycotic solution and 1 ml of L-glutamine. Mix in a beaker in sterile conditions.
Some precipitation can occur during the preparation (see: Troubleshooting).
d) Flame the neck of a sterile bottle over alcohol burner and filter the complete medium into this bottle throughout a single use, 0.20 μm filter using a 20 ml syringe.
e) Flame the neck of the cap, close the bottle, mark necessary details and store at +4°C.

It is recommendable to complete the required amount of medium one day before setting up the culture to reduce the risk of contamination during fish samples manipulation. The complete medium can be stored until the next day.

3. Sampling the fin clips

a) Switch on the UV light in flow box at least two hours before starting the work. Wash hands carefully with antiseptic liquid. Wipe the working area in flow-box with paper tissue wetted in 70% ethanol.

b) Place the culture flasks into flow-box and check whether all necessary items and tools are inside.
c) Place three (about 5 cm in diameter) sterile Petri dishes per sample into the flow-box. Fill them up with 2 ml of complete medium each.
d) Place three additional Petri dishes of similar size in the box. Fill up two Petri dishes with 70% ethanol and one with 1% solution of sodium hypochlorite. This set is used for all the samples according to sterilization method A.

Sterilization method A

a) In the laboratory (possibly close to the flow-box) cut the piece(s) of pectoral or anal fin, place it (them) in sterile Petri dish into few drops of physiological saline solution or complete media until all fin clips are sampled. Move the Petri dishes containing the samples into the flow-box. (In the meanwhile the fish specimens used for sampling are recovered into appropriate medical baths to prevent mycotic infection).
b) Sterilize the tips of forceps (by dipping into 70% ethanol, and than carefully letting the ethanol burn out over the flame).
c) Using the sterilized forceps, transfer the samples into the Petri dish with 70% ethanol (for a few seconds), and immediately after into the Petri dish with 1% solution of sodium hypochlorite (20 s), then dip it into the second Petri dish with ethanol (10 s).
   Check carefully the duration of the treatments.
d) Put the fin clip with sterilized forceps in the first Petri dish with complete medium (first wash).
e) Repeat step d) for another wash in the second Petri dish with complete medium.
f) Repeat step d) for the final wash of the fin clip(s) in the third Petri dish with complete medium.

The samples can be left in the last wash of culture medium for around one hour (until all samples are prepared).

Sterilization method B
(modified from Fontana et al. 1995).

This procedure is much less traumatic for the tissues and reduces the timelag for the beginning of the cell growth from the explants. Unless the fish came from highly polluted waters (needing a more powerful sterilization of the fin clips), this sterilization methodology can shorten the procedure from 3 weeks to 10–12 days, including sub-culture.

Place pieces of fins in a sterile Eppendorf tube containing a solution of sterilized PBS with antibiotic/antimycotic (GIBCO 15240-062), for around two hr at room temperature.

4. Setting up the culture

Sterilize the forceps as already described, before each use. Always chill the tips in the medium.

a) Cut each fin clip mantained in the third wash of complete medium or in a sterile Eppendorf tube (according to the sterilization method used) into four pieces, at best along and not perpendicular to fin rays, using sterilized scissors.
b) Using forceps and a bacteriological inoculating loop, place the fin pieces on the bottom of empty cultivation flasks. Flame the neck of each flask over an alcohol burner before closing it. Keep flasks in upright position for about 30 minutes.
c) Place the flasks horizontally and drop 0.7 ml of medium into each flask in order to slightly wet the pieces of fins, so that they do not detach from flask bottom; disperse the medium to the margins of the flask using a syringe and then carefully move the bottles in order to wet the pieces of fins.

Use a new needle per flask. Always flame the neck of flask before closing.

d) Carefully place all flasks into a thermostat at 28.5°C. Let incubate for 24 hours.
e) Next day complete the amount of medium to final volume of 3 ml in aseptic conditions.
f) When the cells appear around the explant, change the medium every 4 days by using a laboratory suction pump and a sterile pipette until the outgrowth of fibroblasts is visible.
   Check daily under inverted microscope for cell growth and eventual contamination.
g) Cell culture can be harvested for karyotyping or sub-culture when cell outgrowth reaches some 1 cm in diameter around initial piece of fin.

5. Harvesting cells and karyotyping

A. Karyotyping from primary cell cultures

Keep all the procedures aseptic

a) Add 1–2 drops of 0.1% colchicine to each flask and let incubate for 4 hr.
b) Remove the whole medium by suction pump and sterile pipette and wash carefully the cell culture with 2ml of PBS, twice.

c) Detach cells from the bottom of flask by adding 1 ml PBS and 0.7 ml trypsin solution. After at most 2 min, shake the flask rather rigorously and stop the treatment by adding 3 ml of medium.
Check under inverted microscope whether most cells have separated from each other. If so, add 10 ml of medium. If not, shake the flask again.
d) Transfer the content of each flask to a 15 ml tube, centrifuge at 1200 RPM for 8 min.
e) Remove supernatant and add 5 ml of pre-warmed (37°C) 0.075 M KCl. Let it hypotonize for 10 min.
f) Prefix by adding 3–5 drops of freshly prepared 3:1 methanol : acetic acid to each tube, centrifuge at 1200 RPM for 10 min.
g) Fix thrice and spread the cells suspensions onto slides according to conventional protocols for chromosome preparations.

B. Karyotyping from sub-cultures

Keeping all the procedures aseptic.
From a) to d): as described above for A.

e) Remove the supernatant until 1 ml, mix the cell pellet and add 9 ml of media.
f) Put this cell suspension to a bigger (100 cm^2) culture flask.
g) Within two to four days of subculture, harvest the cells and proceed with karyotyping as already described in A.

Sub-culture can be also be performed by adding 3 ml of medium to the original flask; there are always some cells left that can continue to grow.

Troubleshooting

Initiation of primary cell cultures is the most critical step of the procedure. Several problems can affect the attainment of cell monolayer growth. Here are some of the most common troubles.

a) Bacterial or mycotic contamination. To avoid contamination, training to get routine skills in aseptic procedures (washing hands, preparing and sterilization of working place, setting, manipulation and exchange medium, exchange of needles, flaming necks of flasks, etc.) is recommended. A good organization of the working place is necessary.
b) Sterilization washes: small fins require shorter washes than bigger fin clips. Too short sterilization washes result in contamination; too long washes however can kill suitable cells in fin tissues. This parameter should be tested for each studied group/species.
c) The serum completing the medium. Fetal calf serum (FCS) is recommended in all cell culture protocols and it is also used for the fibroblast cultures described here. In some cases, however, autologous serum prepared from

the species under study can be tested for better results. Different batches of FCS even of the same supplier can have different effects on success of culture. Note that the quality of FCS (even from well-known trademarks) is variable and some batches of FCS may not provide good results.

d) Precipitation during the preparation of the complete medium: if coagulation occurs and cloudy precipitates appear in the medium, centrifuge it at high-speed (e.g., 1200 RPM in Eppendorf centrifuge 5804 R) for 30 min, remove the supernatant, filter using aseptic procedures and store.
e) Incorrect attachment of the explants to the bottom of the flask: pay particular care during handling to avoid unnecessary movements of the flasks that can affect the attachment.
f) Modification in the pH: it can be easily checked due to the pH indicator included in the medium. Changes in color of the medium to pink and yellow indicate too high and too low pH values, respectively, generally due to a source of contamination.
g) Temperature: the standard of 28.5ºC in this protocol was experimentally found as the best for cyprinids. For other groups/species different temperatures should be tested.

Laboratory Set up

Equipment

Major

UV laminar flow-box; refrigerator (4°C); freezer; refrigerated centrifuge, low and high speed; mobile rotor; incubator; fast release pipette pump; alcohol or gas burner.

Minor (dispensable sterile or sterilizable)

Plastic culture flasks (25 cm^2 and 100 cm^2); plastic Petri dishes (diameter 55 mm); 15 ml conical tubes with screw caps; fine scissors; curved and strait forceps; inoculating loops; Pasteur pipettes; syringe driven filter unit 0.22 µm; syringes (20 ml, 10 ml, 2 ml); needles 12/5″; micropipette (10–100 µl) and micropipette tips; graduated cylinders (100 ml); Durex bottle (100 ml); beaker (250 ml); timer.

Biochemicals and reagents

Antiseptic soap or solution; Medium L-15 (Leibowitz); fetal calf serum (FCS), stored frozen in 20 ml sterile aliquots; L-glutamine, stored frozen in 1 ml aliquots; antibiotic/antimycotic solution, stored frozen in 0.5 ml aliquots

(Sigma-Aldrich); trypsin, lyophylized; sodium hypochlorite; phosphate buffered saline (PBS); colchicine; methanol; ethanol; glacial acetic acid.

Stock and working solutions

Trypsin stock solution at 10% in sterile distilled water, stored frozen in 0.5 ml aliquots.

Trypsin working solution: in a sterile beaker, dissolve one aliquot of stock solution in 4.5 ml of sterile distilled water; mix and use 0.7 ml of this solution per flask.

References

Amemiya, C.T., J.W. Bickham and J.R. Gold. 1984. A cell culture technique for chromosome preparation in cyprinid fishes. Copeia 232–235.

Fontana, F., M. Lanfredi, R. Rossi, P. Bronzi and G. Arlati. 1995. Established cell lines from three sturgeon species. The Sturgeon Quarterly 3(4): 6.

Ojima, Y. 1978. Preparation of cell-cultures for chromosome studies in fishes. Proc. Japan. Acad. 54, Ser. B: 116–120.

Ojima, Y. 1982. Methods in fish cytogenetics. The Nucleus 25(1,2): 1–7.

11

Cephalic Kidney and Spleen Cell Culture in Antarctic Teleosts

O. Rey,[1,*] *A. d'Hont,*[2] *J.-P. Coutanceau,*[3,a] *E. Pisano,*[4] *S. Chilmonczyk*[5] and *C. Ozouf-Costaz*[3,b]

[1]Station d'Ecologie Expérimentale du CNRS à Moulis, USR 2936, 09200 Moulis, France.
e-mail: olivier.rey.1@gmail.com

[2]Cirad, UMR AGAP Avenue Agropolis, 34398 Montpellier cedex 5, France.
e-mail: dhont@cirad.fr

[3]Institut de Biologie Paris-Seine, UMR "Evolution Paris-Seine", Case 5, 7 quai Saint Bernard, 75252 Paris Cedex 05.
[a]e-mail: coutance@mnhn.fr
[b]e-mail: ozouf@mnhn.fr

[4]DISTAV–Dipartimento di Scienze della Terra, dell'Ambiente e della Vita, University of Genova, Viale Benedetto XV, 5 16132 Genova, Italy.
e-mail: pisano@unige.it

[5]Laboratoire de virologie et immunologie moléculaire, INRA, 78352, Jouy en Josas Cedex, Grance.
e-mail: stefan.chilmon@free.fr
*Corresponding author

Principles

In adult teleost fishes, hematopoiesis mainly takes place in the cephalic kidney where stem cells proliferate to give rise to the various blood cell lineages. In addition, the spleen, the major lymphoid organ, also retains the capability to originate lymphocytes from lymphoid cell precursors. Since cephalic kidneys and the spleen regularly produce dividing cells, their tissues are commonly used for preparing fish chromosomes through direct *in vivo* methods (see Chapter 3, 4, 6). However fish species adapted to Antarctic waters have a low metabolism and can show complete aplasia, especially after the reproductive season, so the usual *in vivo* chromosome preparation techniques often do not produce good results. After some unsuccessful attempts to develop an effective method based on lymphocyte culture (as described in Chapter 9), in these species, we developed a protocol for preparing chromosomes by cell culture from the cephalic kidney and spleen. This protocol has been performed in a field laboratory in Terre Adelie, and has also been tested on board a research vessel in the same sector of the Southern Ocean. The method always produced a higher mitotic index than the direct *in vivo* methods in the polar species tested. A suspension of cephalic kidney and/or spleen cells is prepared and cultured in L-15 Leibovitz's culture medium, without bicarbonate, supplemented with L-glutamine, fetal calf serum, lectins and antibiotics, for a period of up to one week at 0° to +2°C. Colchicine is added 6 hrs prior to harvesting cells. This is followed by a hypotonic treatment (1 hr at +2°C) and conventional steps of fixation. The most efficient lectins for stimulating cell division in these fishes are concanavalin A and pokeweed mitogen, used separately or combined together. Phyto-hemaglutinin (PHA) and lipopolysaccharide of *Escherichia coli* (LPS) did not provide satisfying results.

Applications and Limitations

This protocol has been successfully tested on a large spectrum of notothenioid teleost fishes (genera *Dissostichus, Notothenia, Trematomus, Gymnodraco, Pogonophryne, Chionodraco,* etc.), which live in coastal Antarctic regions at sub-zero temperatures all around the year (–1.8°C), over a wide depth range (from shallow water to 1500 m depth) (see examples of results in Fig. 1). The protocol has a higher rate of success than direct *in vivo* methods with these fishes, thus avoiding the sacrifice of too many specimens. It is also useful for species that cannot be maintained in aquariums for a long time, or which quickly die after capture. All species can produce a high rate of dividing cells (up to 150 analysable metaphases per slide in a single drop of cell suspension) if cultivated in optimal conditions. Cephalic kidney generally produces a higher number of dividing cells than the spleen. For small fish, it is often necessary to mix the spleen and cephalic kidney tissues together, in order to have a sufficient quantity of cells for the volume of a culture flask. For large fish, we recommend preparing two culture flasks, from the right and left cephalic kidney separately, since the tissue can show a different mitotic index.

We expect that this protocol of "rapid cell culture" could also be applicable to sub-Antarctic notothenioids, or to other polar fish, with similar cold-adapted physiology, providing some adjustments to the temperature of the culture (generally 2°C to 4°C more than the optimum temperature of the fish), and to the duration of culture (5, 6, 7 days).

The main limitation to the procedure is the batch of serum, which is used to enrich the culture medium (and is its most expensive component). Sera cannot be tested prior to field work and while some batches can be extremely efficient in boosting cell culture, some are incompatible and not tolerated by fish cells. The best would be to test several different batches on a common

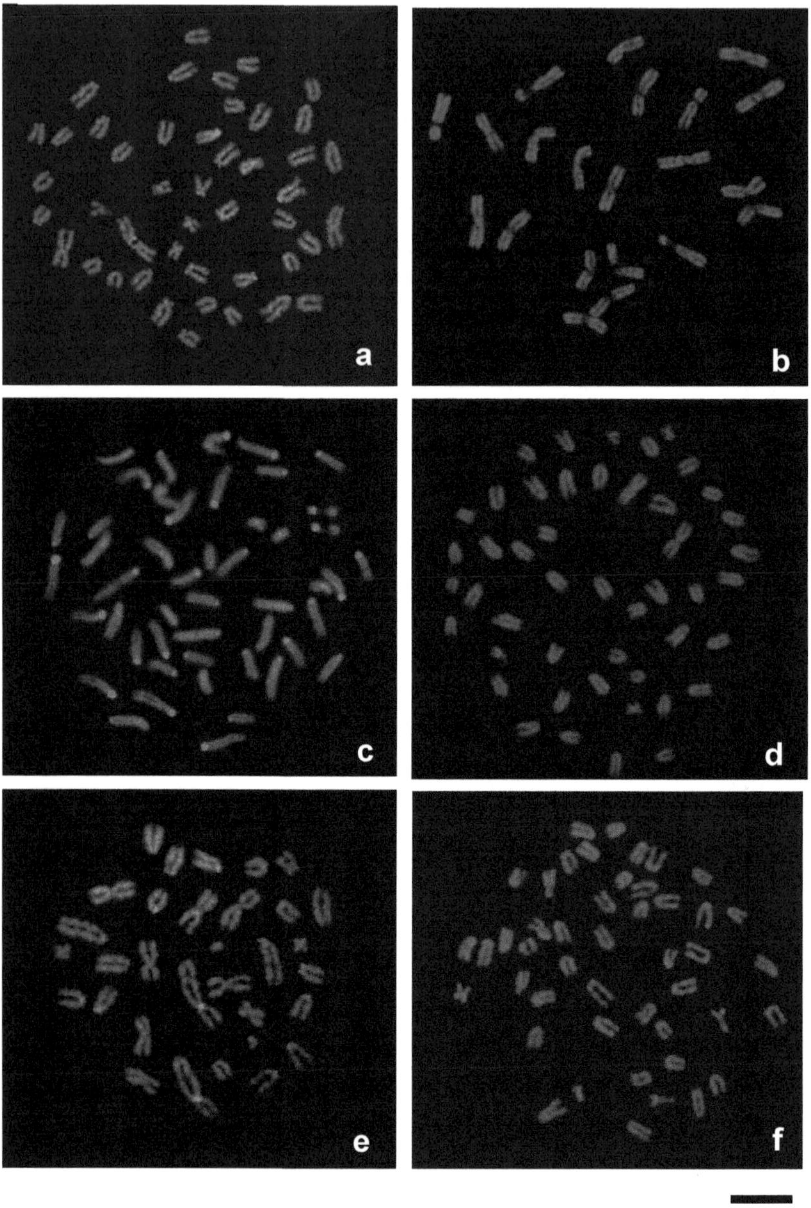

Figure 1. Metaphase plates from kidney and spleen cell culture in Antarctic notothenioid fish species: (a) *Trematomus hansoni*; (b) *Notothenia coriiceps*; (c) *Gymnodraco acuticeps*; (d) *Trematomus newnesi*; (e) *Trematomus pennellii*; (f) *Chionodraco hamatus* (DAPI staining). Scale bar: 10 µm.

Color image of this figure appears in the color plate section at the end of the book.

species before using them on rare species. When a particular batch is found suitable in one species, it is generally efficient with other species, if these are closely related.

Protocol

Preparation of working space and tools

Work as much as possible in sterile conditions. If possible, manipulate under a laminar flow box, or at least, close to the flame of a Bunsen burner. Decontaminate the working space (benches, equipment, dissecting tools) with UV light and/or with 70% ethanol.

Prior to dissection, prepare a suitable quantity of the complete medium in a sterile flask.

Prepare as many sterile Petri dishes, cell strainers and culture flasks as fish tissues to be cultivated.

In a polystyrene box, spread a layer of crushed ice. Cover with a triple layer of filter paper and a layer of aluminium foil. Place Petri dishes on top and distribute with a sterile pipette, 1 ml of complete culture medium and a cell strainer in each dish. The mesh of the cell strainer must be in contact with the liquid, but not flooded. Cover the strainers with the Petri dish lids. Petri dishes must be maintained in such cool conditions during the whole manipulation and cell dissociation, but cells must never freeze.

Preparation of culture flasks (25 cm^2)

In each flask and with sterile pipettes or syringes, place:

- 8 ml complete medium
- 1 ml mitogen 1x (concanavalin A, pokeweed, or both combined together)

Store on crushed ice beside the Petri dishes or at +4°C before dissection.

Dissection

Use sterile dissecting tools.

Anaesthetize the fish (according to Machin 2001).

Spleen and kidney tissues can be cultivated together or separately. Not more than a total amount of 1 g (c.a. the size of a corn seed) must be cultivated per flask.

Wash the fish abdomen with 70% ethanol or chlorexidin solution.

Using sharp scissors, open the abdominal cavity from anus to chin, cut the intestine backwards, cut the oesophagus and carefully remove the whole digestive tract. Also remove the gonads, if they are well developed. Remove a spleen fragment (or the whole spleen if the fish is small) and transfer it on the cell strainer. Partly cover with the Petri dish lid.

Briefly wipe the abdominal cavity with soft tissue and chlorexidin and take off the peritoneum to clear the cephalic kidney. Remove a piece of this from its most anterior part and transfer it on the cell strainer.

Cell dissociation

Dissociate the tissue on the cell strainer with the rounded part of curved forceps. Do not use the forceps tips, to avoid getting strainer nylon particles into the cell suspension. Slightly lean the Petri dish and collect the suspension with a sterile Pasteur pipette on the edge; eventually rinse the strainer two or three times before transferring the suspension into the culture flask and screw-close the lid.

Cell culture

Transfer the flask to a refrigerator at 0°C to +2°C, and keep it closed, in a flat position, neck upwards, in the dark, for a period of 6.5 to 7.5 days. Maintain the temperature constant. Once per day, gently shake the flask in order to re-suspend the cells. Any change of colour in the medium (to pink or yellow) may indicate a contamination, possibly leading to cell destruction and the absence of dividing cells at the harvesting step.

Colchicine treatment

Six hours prior to cell harvesting, add 20 µl of 0.5% colchicine solution per flask, mix and store again in the refrigerator in a horizontal position.

Cell collection and hypotonic treatment

Gently agitate to re-suspend the cells and transfer the suspension to 15 ml conical tubes.

Centrifuge 10 min at 1500 rpm.

With a Pasteur pipette, remove the supernatant and replace it by 12 ml of 0.075 M KCl.

Store at +4° for one hour and gently re-suspend every ten minutes.

Prefixation

Re-suspend and add 1 ml of ice-cold fixative, drop by drop, to the tube. Gently mix by up and down movement and centrifuge for 10 min at 1000 rpm.

Fixation 1

Remove supernatant, leaving just 2 mm on top of the pellet and homogenize by slight pumping with the Pasteur pipette. Add ice-cold fixative drop by drop and progressively mix between each drop in order to avoid cell agglutinations. Fill up the tube with ice-cold fixative and leave the tube in crushed ice for 10 min. Centrifuge for 7 min at 2000 rpm.

Fixations 2 and 3

Repeat fixation 1 step twice. Reject supernatant and re-fix in 1.5 ml of ice-cold fixative. Maintain in crushed ice.

Slide preparation

Place a clean, dry slide on the bench in a horizontal position. Mix the suspension well with the Pasteur pipette and then allow a single drop of suspension to fall onto the very middle of the slide, from an approximately 2 cm height. The drop should make a regular circle where the differential evaporation of acetic acid and methanol will allow a good spread of nuclei and chromosomes on the slide surface. Slide preparations can be stored at –20°C in airtight boxes for a very long time (up to 10 years, probably more).

Cell suspensions can also be stored at –20°C in well-sealed conical tubes filled with fresh fixative.

Check the spreading quality (cell dispersion) and mitotic index under a microscope, with a phase contrast microscope (magnification x 10). In case of insufficient hypotonic treatment, see "Troubleshooting" section.

Troubleshooting

The temperature of the refrigerator in which the cells are to be cultivated must be tightly controlled. Below 0°C, cells start to die. The refrigerator should remain closed as much as possible to avoid temperature fluctuations and light exposure. If the door has to be frequently opened, culture flasks must be put in a closed isolating polystyrene box. On board ships, flasks must not be shaken or knocked during culture: they can be attached to an anti-rolling plate with scotch tape.

Sterile conditions are practically impossible to attain in field laboratories, but in most cases, contaminations in the culture flasks can be controlled or limited by the antibiotic-antimycotic components.

Another cause of failure is that the kidney hematopietic tissue is sometimes difficult to identify and dissect since blood clots which form around it can get mixed up with it and collected, instead. Sometimes the dissection is performed too slowly, or at a too cold or too warm a temperature, which can damage the cells and prevent them from recovering in the culture medium. Cells must never be maintained at subzero temperature after dissection, because they are no longer protected by antifreeze compounds as in the *in vivo* tissue.

In Antarctica, the ambient air can be very dry. In such conditions, cell suspensions dry too fast when they are dropped onto slides, resulting in insufficient chromosome scattering, even if a correct hypotonic treatment has been applied. This problem can be avoided by preparing the slides in a hood and close to a humidifier.

Some other culture media have been tried with cephalic kidney and spleen cells in similar culture conditions: RPMI 1640 (Gibco) does not produce any dividing cells; PB-MAX karyotyping medium (Life Technologies) supplemented with concanavalin A produced a very small amount of dividing cells.

Laboratory set up

Equipment

Major

UV laminar flow box (optional); refrigerator; refrigerated centrifuge; low speed; mobile rotor; fast release pipette pump; gas burner.

Minor

Sterile plastic culture flasks (25 cm^2); polystyrene sterile petri dishes diameter 55 mm; sterile syringes and pipettes; sterile Falcon cell strainers, 70 µm; micropipettes and sterile micropipette tips; Eppendorf tubes, dissecting tools; Pasteur pipettes; 15 ml conical tubes; beakers and cylinders.

Biochemicals and Reagents

L-15 Leibovitz's with L-glutamine, without bicarbonate for culture without CO_2 (Gibco 11415); foetal calf serum for cell culture; antibiotic-antimycotic (Gibco 0343); colchicine; concanavalin A (Sigma C-5275, 5 mg, lyophylized, –20°C); Pokeweed (Lectin from *Phytolacca americana*, Sigma L-9379, 5 mg, +4°C); KCl, 0.075 M; chlorexidin; absolute ethanol; glacial acetic acid.

Stock and working solutions

Complete culture medium: mix up 79 ml L-15 Leibovitz's medium with L-glutamine, without bicarbonate; 20 ml foetal calf serum; 1 ml antibiotic antimycotic.

N.B. Store for not more than 15 days at +4°C, or deep freeze. Add 1 ml of glutamax-1 100X (Gibco 0312) every four days, because L-glutamine is unstable at physiological pH in liquid media.

Mitogens 1x:

Optional:

- Concanavalin A
  Stock solution, 1 mg/ml: dilute 5 mg concanavalin A in 5 ml of sterile L-15 medium.

  Working solution, 1x: dilute 1 ml of stock solution in 9 ml sterile L-15 medium and distribute 1 ml aliquots in sterile Eppendorf tubes. Store at –20°C and thaw just before use.

- Pokeweed mitogen
  Stock solution, 1 mg/ml: dilute 5 mg Pokeweed in 5 ml of sterile L-15 medium.

  Working solution, 1x: Dilute 1 ml of stock solution in 9 ml sterile L-15 medium and distribute 1 ml aliquots in sterile Eppendorf tubes. Store at –20°C and thaw just before use.

- Combined concanavalin and pokeweed mitogens, 1x
  Dilute 1 ml of concanavalin A stock solution in 4 ml of L-15 medium. Aliquot in sterile Eppendorf tubes, 0.5 ml per tube.

  Dilute 1 ml of pokeweed mitogen stock solution in 4 ml of L-15 medium. Add 0.5 ml aliquots of this solution in the Eppendorf tubes already containing 0.5 ml of concanavalin A.

  Deep-freeze at –20°C and thaw just before use.

Acknowledgements

This protocol could be obtained thanks to financial support of CNRS, MNHN, ICOTA and REVOLTA programmes of the Paul Emile Victor Institute (IPEV), and CEAMARC-CAML programme. We are very grateful to Stéphanie Pavoine, Julien Lanshere, Thibaut Nebout and Elsa Freschet for their technical help during field work and to P.A. Hulley for kindly reviewing this manuscript.

Reference

Machin, K.L. 2001. Fish, amphibian and reptile analgesia. Vet. Clin. North Am. Exot. Anim. Pract. 4(1): 19–33.

12

Meiotic Surface-Spread and Immunodetection for Synaptonemal Complex Proteins in Teleost Fish

C. Araya-Jaime,[a,*] *E. Alves-Serrano, C. Oliveira* and *F. Foresti*

Departamento de Morfologia, Instituto de Biociências, Universidade Estadual Paulista, Distrito de Rubião Junior, s/n, 18618-970, Botucatu, SP, Brazil.
[a]e-mail: carayaj_uls@yahoo.es
*Corresponding author

Principles

Studies of meiotic chromosomes provide important information about the behavior and the formation of sex chromosomes, origin and evolution of supernumerary chromosomes, and chromosomal rearrangements involved in the karyotype evolution. The protocol of meiotic chromosome preparations for synaptonemal complex analysis in teleost fish presented in this chapter is based on methods described by Van Eenennaam et al. (1998) and Ocalewicz et al. (2009) with modifications. The implementation of this protocol allows the application of several techniques, such as chromosome banding, FISH and immunodetection, once the chromatin of the bivalent chromosomes is preserved (for example see Fig. 1). It also allows independently identifying the central or the lateral elements of the synaptonemal complex by using different immuno-fluorescent probes. The greatest challenge for the success of this technique is the need to understand the reproductive cycle and the biology of the species studied to know at which maturity stage it is possible to obtain the synaptonemal complex.

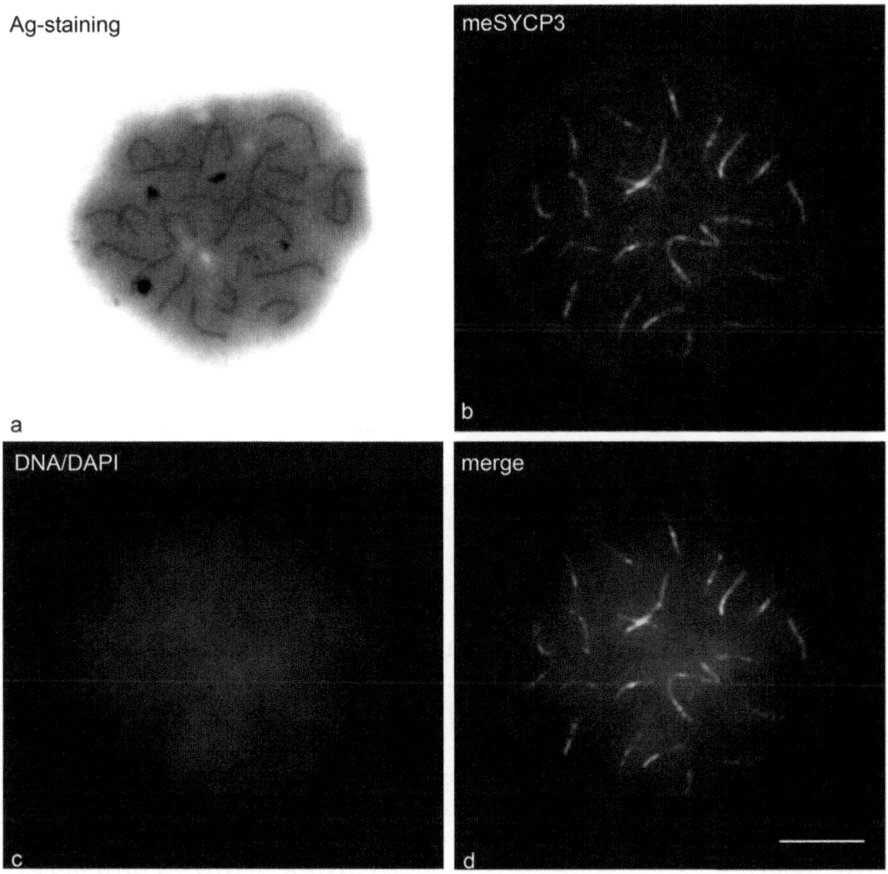

Figure 1. Photomicrograph of an *Oreochromis niloticus* pachytene spermatocyte stained by silver nitrate (a). The stained SC elements are covered by the chromatin of bivalents. In (b), pachytene spermatocytes stained with the anti-medaka SYCP3 antibody. SCs are selectively labeled. The same cell was stained with the DNA-specific fluorochrome DAPI, showing the chromatin only (c). In (d) association of the two images. Scale bar 10 µm.

Color image of this figure appears in the color plate section at the end of the book.

Applications and Limitations

Analysis of the synaptonemal complex (SC) in fish enables further knowledge about the dynamics and evolution of sex chromosomes, studies about the inheritance and origin of supernumerary chromosomes, as well as identification of chromosomal rearrangements. Besides the meiotic studies in fish are still restricted to a few species, interesting information was obtained in studies on *Danio rerio* (Wallace and Wallace 2003), *Acipenser transmontanus* (Van Eenennaam et al. 1998), *Oryzias latipes* (Iwai et al. 2006) *Oncorhynchus mykiss* (Oliveira et al. 1995) and *Oreochromis niloticus* (Ocalewicz et al. 2009), among others.

The main difficulty in obtaining good preparations of (SC), is the need for obtaining dividing cells in the pachytene stage. This requires the determination of the exact gonad maturation phase of the studied organism. In fish for example, there are species that breed throughout the year, others once a year, or at certain times of the year (Schulz et al. 2010). Different patterns of reproductive behavior, and environmental changes can directly influence the maturation period. Therefore, it is necessary to understand the reproductive cycle of the species studied to know the right time and stage of maturity at which it is possible to obtain spermatocytes at the pachytene stage and consequently, the SC.

Our protocol allows the use of specific antibodies for fish, as well as commercial antibodies derived from mammals, once performed a pretreatment with citrate buffer, leaving the sites for complex of proteins most exposed to binding to the antibodies. Positive results were obtained using antibodies for SYCP3 protein found in the lateral elements of the SC in different fish species, such as *Oreochromis niloticus*, *Eigenmannia* sp., *Astyanax paranae*, and *Moenkhausia sanctaefilomenae*, demonstrating that this protocol can be used in various teleost fish groups.

The preparations obtained with this protocol can be used in association with other cytogenetic techniques, for example silver nitrate staining (Fig. 1a), immunostaining (Figs. 1b, c, d) and localization of DNA markers by fluorescence *in situ* hybridization combined with immunostaining, in order to visualize simultaneously the behavior of chromosomes, unpaired regions, as well as the location of molecular markers. It can also be used in the application of the whole chromosomes painting technique during pachytene (Ocalewicz et al. 2009).

Protocol

1. **Surface-spreading Protocol**, modified from Van Eenennaam et al. (1998) and Ocalewicz et al. (2009).
   a) Sacrifice individuals with overdose of benzocaine.
   b) Remove the gonads (testis or ovarians) and immerse in cold Hank's solution (Sigma), the volume of this saline solution depends on the size of the gonads.
   c) Mince tissue to a fine cell suspension (as described in Chapter 3 or 11).
   d) Transfer the cell suspension to a 1.5 ml tube and decant for 30 min.
   e) Place a drop of this suspension (20 μl approximately) in the middle of a clean, dry slide, avoiding the removal of cell debris.
   f) Add 2 drops of 0.2 M sucrose, and 2 drops of Triton X-100 0.2%, all the solutions adjust to pH 8.5 with sodium tetraborate 0.1 M.

g) Incubate for 4 minutes preparations with testis cell suspension or 7 min preparations with ovarian cell suspension.
h) Fix the cell suspension by adding 10 drops of 4% paraformaldehyde (pH 8.5 with sodium tetraborate 0.1 M) for 15 min.
i) With the help of an absorbent paper, remove excess of liquid, and letit dry at room temperature for 3 hrs.
j) Wash 3 times with Photo Flo 200 (Kodak) 0.08%, and dry in a vertical position.
k) Assess the quality of the preparations by staining with silver nitrate according to Howell and Black 1980.
l) Store preparations at –20°C in the deep freezer.

2. **Immunostaining,** modified from Berrios et al. (2010).
a) Wash the slide three times in PBS 1x (prepared from the PBS 10x), 5 min, do not dry.
b) Antigen recovery*, immerse the slides in 0.01 M citrate buffer pH 6, preheated to 95°C for 20 min.
c) Allow to cool at room temperature for 20 min.
d) Wash three times in PBS 1x, 2 min each.
e) Prepare the primary antibody solution (1/100) in PBS 1x, drooped 25 µl per slide, with coverslip.
f) Incubate with primary antibody for 1.5 hrs at 37°C in a humid chamber*.
g) Wash three times in PBS 1x, 5 min each.
h) Incubate with secondary antibody (1/100 in PBS 1x), 25 µl per slide, with coverslip, for 40 min at 37°C.
i) Wash three times PBS 1x, 5 min each.
j) Mount preparations under coverslip whithVectashield/DAPI.

* The times and temperature for primary antibody incubation can be modified, 3 hrs at room temperature or overnight at 4°C.

Troubleshooting

1. The density of the cell suspension should be checked, because if it is too dense it can compromise the quality of the cell spreading.
2. On cold days, drying the slide after fixing can exceed the indicated time (3 hours). It is important to wait until the slide is completely dry.

Laboratory set up

Equipment

Major

Incubator; pHmeter; Freezer; Refrigerator.

Minor

Humid chamber; Petri dish (diameter 55 mm); 1.5 conical tube; tweezers; Pasteur pipettes; micropipette; micropipette tips; slides; coverslips (24 x 50 mm); absorbent paper; and timer.

Biochemicals and reagents

Benzocaine (Sigma); Hank's salt solution (Sigma); 0.01 M Citrate buffer pH 6; 0.2 M Sucrose; 0.1 M Sodium tertraborate; Triton X-100 0.2%; paraformaldehyde 4%; Photo Flo 200 (Kodak) 0.08%; Phosphate buffer saline (PBS) 1x; Primary antibody solution GuineaPig-AntiSCPY3-Medaka (Iwai et al. 2006); Secondary antibody solution AntiGuineaPig-FITC (Santa Cruz Biotechnology), Vectashield + Dapi (Vector Labs).

Stock and working solutions

Buffer Citrate 0.01 M
2.94 g Tri-sodium citrate (Merck, PM = 294,10 g/mol)
1000 mL distilled water
0.05% Triton^{R}-X100 (USB^{TM}# 9002-93-1)
Sucrose 0.2 M
3.42 g Sucrose (USB^{TM}, PM= 342,30 g/mol)
50 mL distilled water
pH 8.5 with sodium tetraborate 0.1 M.
Triton X-100 0.2%
30 µL Triton^{R} X-100 (USB^{TM}# 9002-93-1)
15 mL distilled water
pH 8.5 with sodium tetraborate 0.1 M
Sodium tetraborate 0.1 M
3,813 g Sodium tetraborate + 10 H_2O ($Na_2B_4O_7$) (Merck^{TM})
100 mL distilled water.

References

Berrios, S., M. Manterola, Z. Prieto, J. López- Fenner, J. Page and R. Fernández-Donoso. 2010. Model of chromosome associations in *Mus domesticus* spermatocytes. Biol. Res. 43: 275–295.

Howell, W.M. and D.A. Black. 1980. Controlled silver staining of nucleolus organizer regions with a protective colloidal developer: A 1-step method. Experientia 36: 1014–1015.

Iwai, T., A. Yoshi, T. Yokoto and C. Sakai. 2006. Structural components of the synaptonemal complex, SYCP1 and SYCP3, in the medaka fish *Oryzias latipes*. Exp. Cell Res. 312: 2528–2537.

Ocalewicz, K., J.C. Mota-Velasco, R. Campos-Ramos and D.J. Penman. 2009. FISH and DAPI staining of the synaptonemal complex of the Nile tilapia (*Oreochromis niloticus*) allow

orientation of the unpaired region of bivalent 1 observed during early pachytene. Chrom. Res. 17: 773–782.
Oliveira, C., F. Foresti, M.G. Rigolino and Y.A. Tabata. 1995. Synaptonemal complex analysis in spermatocytes and oocytes of rainbow trout, *Oncorhynchus mykiss* (Pisces, Salmonidae): the process of autosome and sex chromosome synapsis. Chrom. Res. 3: 182–190.
Schulz, R.W., L.R. França, J.J. Lareyre, F. LeGac, H. Chiarini-Garcia, R.H. Nobrega and T. Miura. 2010. Spermatogenesis in fish. Gen Comp. Endocr. 165: 390–411.
Van Eenennaam, A.L., J.D. Murray and J.F. Medrano. 1998. Synaptonemal complex analysis in spermatocytes of white sturgeon, *Acipenser transmontanus* Richardson (Pisces, Acipenseridae), a fish with a very high chromosome number. Genome 41: 51–61.
Wallace, B.M. and H. Wallace. 2003. Synaptonemal complex karyotype of zebrafish. Heredity 90: 136–140.

13

C-banding

S. Salvadori,[a,*] *E. Coluccia*[b] and *A.M. Deiana*[c]

Dipartimento di Scienze della Vita e dell'Ambiente, Università di Cagliari, Via T. Fiorelli 1, 09126 Cagliari, Italy.
[a]e-mail: salvador@unica.it
[b]e-mail: coluccia@unica.it
[c]e-mail: adeiana@unica.it
*Corresponding author

Principles

C-banding is a staining technique specific for most constitutive heterochromatin (in some cases there are heterochromatic regions not demonstrated by C-banding methods) but not for facultative heterochromatin (Sumner 1990; 2003). C-positive regions are regarded as sites of highly and moderately repetitive DNA, even in the absence of direct evidence (Sumner 2003). The mechanism of C-banding involves the preferential removal of DNA from non-C banded chromosomal regions by depurination of DNA, hydrolysis of depurinated DNA and extraction of hydrolysed DNA (Holmquist 1979; Sumner 2003). The DNA of C-banded regions is more difficult to extract probably because of a slower depurination of the DNA of condensed chromatin and because of a stronger binding of DNA to proteins (Sumner 1990).

Applications and Limitations

C-bands are detectable in any eukaryote taxon. Generally the size and position of the C-bands are very good chromosome markers in karyological studies and are particularly useful in karyotyping taxa (like fishes) in which G-, R- and Q bands cannot normally be induced (Sumner 1990). In a normal karyotype at least a small band is usually present at the centromere of chromosomes and pericentromeric, telomeric and sometimes intercalary bands are found (Fig. 1). C-bands can be heteromorphic in many species and the heteromorphisms

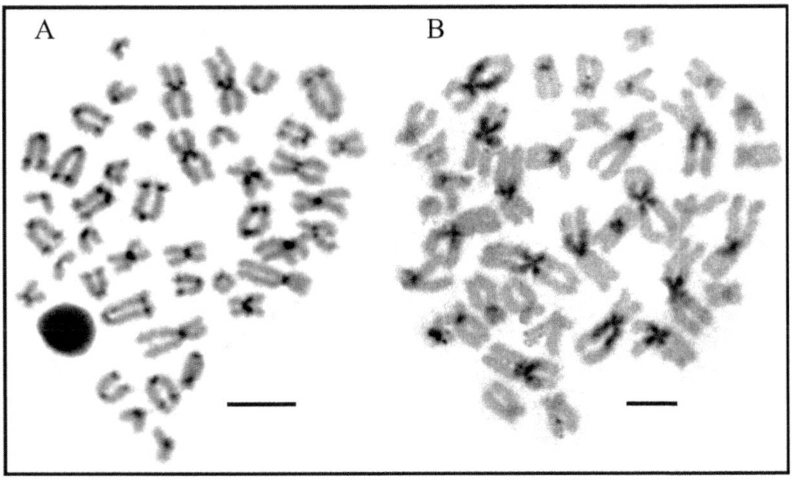

Figure 1. C-banded metaphases from the anguilliform species *Muraena helena* (A) and *Ophisurus serpens* (B).

can be found among individuals and among populations of the same species. Moreover, C-bands are useful markers in karyotype evolutionary studies. In fact, differences in constitutive heterochromatin amount and localization are often found among related species.

Protocol

This BSG (Barium hydroxide/Saline/Giemsa) method is a slight modification of the standard method for producing C-bands (Sumner 1972):

- age the slides for 6 to 12 days (we obtain the best results with 6 to 7 days old slides) or leave them for 1 day in a 60°C incubator.
- immerse in 1 N HCl for 3 min (or alternatively in 0.2 N HCl for 50 min) at room temperature and rinse in distilled water.
- incubate in a freshly prepared saturated (5%) barium hydroxide solution at 50°C for 3 to 4 min. During the preparation of barium hydroxide solution a layer of barium carbonate forms on the surface, it is preferable to remove it with blotting-paper just before and at intervals during the incubation. Alternatively the slides can be incubated in barium hydroxide solution at lower temperatures, e.g., 30°C for 5 min (Schmid et al. 1979).
- rinse briefly in 0.1 N HCl and distilled water in succession, two or three times.
- incubate in 2x SSC, pH 7 at 60°C (50°C for a milder treatment) for one hour.
- rinse in distilled water and air-dry.
- stain in 1:2 Wright's stain in phosphate buffer, pH 6.8 for 10 min or 5% Giemsa, pH 6.8 for 5–15 min.

Troubleshooting

The barium hydroxide treatment is the most critical stage of the procedure: an excessive treatment can cause swelling with pale staining of the chromosomes. The best temperature and time of barium hydroxide treatment should be tested because it could vary from one tissue or species to another.

Laboratory set up

Equipment

Major

Precision balance; pHmeter; water bath; microscope; magnetic stirrer, laboratory oven.

Minor

Microscope slides and coverslips; Graduated measuring cylinders; coplin jars; beakers; timer.

Biochemicals and reagents

Giemsa, 5%; 1 N HCl; 0.1 N HCl; 0.2 N HCl.

Stock and working solutions

$Ba(OH)_2$ solution

Dissolve 5 g $Ba(OH)_2$ in 100 ml of distilled water. The 5% $Ba(OH)_2$ solution is a saturated solution that has to be filtered before the incubation of the slides.

Wright solution

Stock solution: dissolve Wright's powder in methanol (0.25 g in 100 ml) on a magnetic stirrer in the dark; let the solution stabilize for 48 hr at 37°C.

Working solution: dilute Wright solution 1:2 in the phosphate buffer described in the annex at the end of the book at pH 6.8.

References

Holmquist, G.P. 1979. The mechanism of C-banding: depurination and β-elimination. Chromosoma 72: 203–224.

Schmid, M., J. Olert and C. Klett. 1979. Chromosome banding in Amphibia III. Sex chromosomes in *Triturus*. Chromosoma 71: 29–55.

Sumner, A.T. 1972. A simple technique for demonstrating centromeric heterochromatin. Expl. Cell Res. 75: 304–306.

Sumner, A.T. 1990. C-banding and related methods. pp. 39–68. *In*: Sumner, A.T. (ed.). Chromosome Banding. Unwin Hyman Ltd., London.

Sumner, A.T. 2003. Constitutive heterochromatin. pp. 84–96. *In*: Sumner, A.T. (ed.). Chromosomes: Organization and Function. Blackwell Science Ltd., United Kingdom.

14

Sequential Chromosome Banding in Fishes

M. Rábová,[1] *M. Völker,*[2] *Š. Pelikánová*[1] and *P. Ráb*[1,*]

[1]Laboratory of Fish Genetics, Institute of Animal Physiology and Genetics, Academy of Sciences of the Czech Republic, 277 21 Liběchov, Czech Republic.

[2]Institut für Evolutionsbiologie und Ökologie, Universität Bonn, 53121 Bonn, Germany.
e-mail: mv53121@googlemail.com

*Corresponding author: rab@iapg.cas.cz

Principles

Unlike the chromosomes of homeothermic vertebrates, the chromosomes of poikilothermic vertebrates generally do not show euchromatic, serial banding patterns (G-, R-, Q-bands) after treatment with G-banding methods (using trypsin digestion as described by Seabright 1971, or incubation in hot 2x SSC according to Sumner et al. 1971) or with DNA base-specific fluorochromes (Medrano et al. 1988; Sumner 1990). Thus, banding procedures developed to investigate chromosomes of higher vertebrates (especially birds and mammals) are rarely (e.g., Blaxhall 1983; Gold et al. 1990; Yu et al. 1994) reproducible in fish chromosomes. The structural basis of euchromatic or serial banding patterns has been associated with the compartmentalisation of the genome of higher vertebrates into AT-rich and GC-rich isochores, not present in the chromosomes of fishes, most amphibians and reptiles (Medrano et al. 1988; Schmid and Guttenbach 1988; Schmid et al. 1990; Sumner 1990). In any case comparative chromosome banding studies in lower vertebrates are often limited to the determination of 2n, NF and chromosome morphology by using basic staining techniques to describe the conventional traits (such as Giemsa staining), or the pattern of constitutive heterochromatin by using C-banding

(CBG), or the fluorescence banding patterns (by using DNA base-specific fluorochromes). The identification of the nucleolus organising regions (NORs) using GC-specific fluorescent agents (e.g., Mithramycin M–MM, Chromomycin A_3–CMA_3) and impregnation with $AgNO_3$ is also a widely used method in the tool-box of fish cytogenetics. Sequential chromosome banding consists of the sequential application of different staining and banding methods to the same metaphases, thus substantially increasing the information on the chromosome structure. Therefore, sequential banding can be viewed as a methodological compensation for the inapplicability of euchromatic banding techniques in cytogenetics of higher vertebrates.

The principle of sequential banding is based on the evidence that each single treatment has a different action and impact to the chromosomal structure by even partially destroying different components of the chromosomal protein/DNA complexes. Based on the above, the order in which the various staining and banding treatments applied in sequential banding is very critical and the protocol must obviously start with the least destructive procedure.

For example fluorescent banding (based on the preferential binding of fluorochromes, usually antibiotics, to AT- or GC-rich DNA segments, typically heterochromatic regions) is not destructive and the fluorescent dyes can be easily removed from chromosomes by washes with ethanol. Instead, C-banding has a rather destructive impact to the chromosome structure since it is based on a strong (although short) denaturation by NaOH.

Ag-NOR staining, based on the silver impregnation of nucleolar rRNA-protein complexes produces a reversible pattern and the chromosomes can be destained, but their morphology is rather strongly affected by the procedure. Conventional Giemsa staining does not affect the chromosomes and nuclear dye that can be easily removed by washing in fixative. According to their characteristics, the correct steps in a protocol of sequential banding including all the above single treatments, must be the following: fluorescent banding, followed by conventional Giemsa-staining, followed by the application of either CBG or Ag-NOR staining. If all these four steps are required, Ag-NOR staining will be the last. Another combination can be CBG followed by fluorescent banding, thus allowing to visualize more heterochromatic regions (irrespective of base-specificity). This effect is likely due to stronger binding of fluorescent dyes to heterochromatic regions, where the denaturation of DNA was weaker during CBG. As a result, after CBG followed by DAPI (4′,6-diamidino-2-phenylindole) staining, different bands can be detected according to various fluorescent intensity classes (i.e., weak fluorescence indicates euchromatic regions, moderate signals vizualize either GC-rich or heterogenous heterochromatic regions, strong fluorescence indicates AT-rich heterochromatin and complementary GC-rich DNA sequences).

Applications and Limitations

Combined and simultaneous use of fluorescent dyes with different base-specifity together with a counterstain agent, known as CDD-banding fluorescence banding (ChromomycinA$_3$/Distamycin/DAPI) (Schweizer 1980; 1981) to the same metaphase may distinguish different types of heterochromatin. For the first time in fishes, Haaf and Schmid (1984) applied this approach using Distamycin A (DA) as counterstain and the AT-specific DAPI fluorochrome followed by CBG on *Poecilia* chromosomes to detect a ZZ/ZW sex chromosome system. Mayr et al. (1985) combined Ag-NOR staining followed by CDD-banding (CMA$_3$/DA/DAPI—counter-stain enhanced fluorescence) with subsequent DAPI/Actinomycin D (DA) fluorescence on destained slides after Ag impregnation to document chromosomal characteristics of NORs in the karyotype of *Perca fluviatilis*. Later, Mayr et al. (1986) included conventional Giemsa staining on destained slides after Ag-NOR and CDD-bandings as last step. Soon, methyl green as another and more efficient counterstain agent was included into fluorescent banding protocols instead of DA and CBG was combined with DAPI fluorescence (e.g., Sola et al. 1992). This approach using predominantly CMA$_3$ and DAPI was and is the most frequently used in fish cytogenetics. Recently, destained slides that were used for FISH detection of ribosomal gene clusters were sequentially subjected to Ag-NOR staining and fluorescent banding, respectively (e.g., Fujiwara et al. 1998; Martins and Galetti 1999; Gromicho et al. 2005). Combinations of various sequential banding protocols, ideally combined with *in situ* hybridization, are thus an invaluable tool in fish descriptive cytogenetics.

The sequential banding approach provides useful information for a number of applications. For instance the pattern of distribution and base composition of heterochromatin using ante- or post-application of CBG and fluorescence banding can allow the identification of AT-rich heterochromatin chromosomal regions that can be hot spots for recombination (e.g., Shaffer and Lupski 2000). The same approach can allow the identification and characterisation of both homomorphic and heteromorphic sex chromosomes (e.g., Haaf and Schmid 1984). GC-specific fluorescent banding followed by Ag-NOR can be used to identify the sites of major ribosomal gene clusters and their combination with CBG and/or direct visualization after *in situ* hybridization with suitable probes, provides data on the organization of these sites (e.g., Rábová et al. 2001). Sequential banding can identify minor rearrangements, remaining unidentifiable on Giemsa-stained chromosomes (Fig. 1). The sequential CMA$_3$/DAPI banding can be perfomed in one step (CMA$_3$ + DAPI banding in two subsequent staining rounds, followed by recording the chromosomes with the appropriate fluorescence filter combination or in two steps (CMA$_3$ banding—recording the chromosomes/DAPI banding—recording the chromosomes)). Although the one-step procedure is evidently much faster, the two-step protocol seems to provide brighter signals. The choosing between the one or two steps protocols will also depend upon the species, e.g., the chromosomes

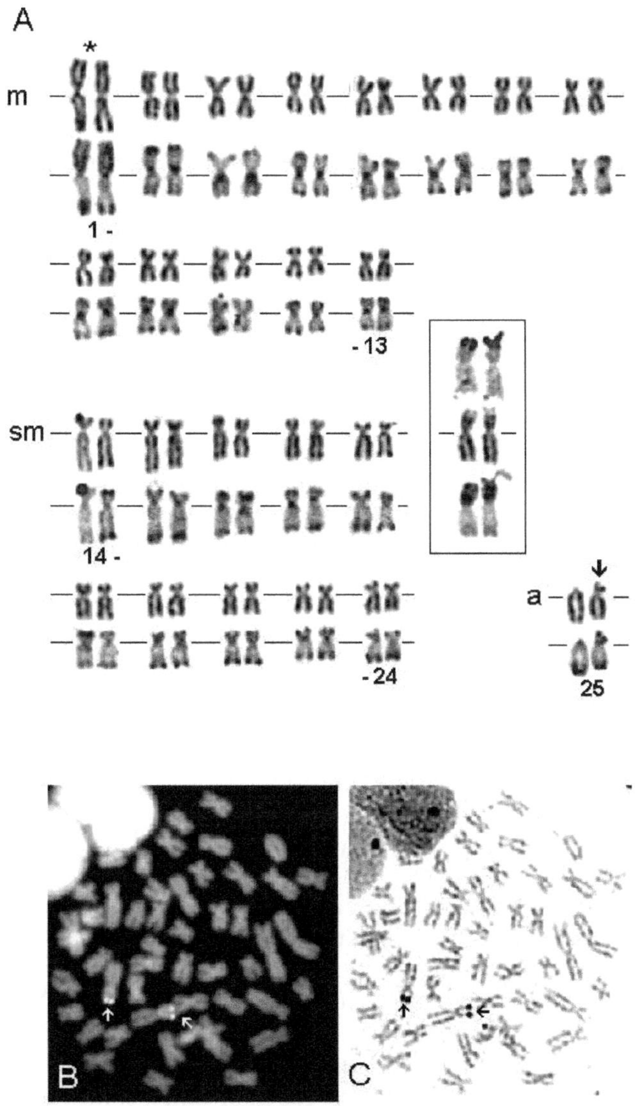

Figure 1. Karyotype (A) of cyprinid fish *Romanogobio kessleri* arranged from sequentially Giemsa-stained (upper row) and C-banded (lower row) chromosomes. Ag-NOR is shown in the frame. Sequentially CMA_3-(B) and Ag-stained (C) chromosomes of the same specimen. The combination of two sequential protocols (Giemsa-staining-C-banding/and Ag-staining, A) and CMA_3-Ag-staining (B,C) shows that NOR-bearing chromosomes has large pericentromeric blocks of C-positive heterochromatin on p arms (framed in A) and that NOR sites are C-, CMA_3- and Ag-positive (arrows in B and C). The specimen is also the homozygous carrier of the entirely heterochromatic p arm of the 1st metacentric pair (asterisk) and the heterozygous carrier of a pericentric inversion in chromosome pair 25 (arrow): such rearrangements are not identifiable in Giemsa-stained chromosomes.

of polyploid cyprinids exhibit very small CMA_3-positive signals and the one-step protocol produces nearly invisible signals.

The successful results of sequential banding largely depends on the quality of the chromosome preparations (degree of chromosome condensation, number of analyzable metaphases per slide) and on karyotypic characteristics of the species/group under study (number and size of chromosomes). Moreover sequential banding requires a reliable system for checking the metaphase coordinates onto the slides because they have to be found and recorded at various steps of sequential banding. As many metaphases as possible per slide have to be analysed because several of them are possibly damaged during the treatments. Usually, a complete successful series of sequential bandings such as CMA_3, DAPI, Giemsa, C-bands and Ag-NORs is very difficult and various combinations of bandings with several slides of the same specimen have to be tested to reach the maximum level of information. Figure 1 provides examples of such banding combinations.

The major limitation of sequential bandings is that, though several or even numerous chromosomal markers and/or segments can be visualized, in many cases the unequivocal identification of chromosomal homologies remains impossible.

Protocols

Counterstain enhanced fluorescence (CDD) protocols are modified combinations of procedures in Mayr et al. (1986), Sola et al. (1992) and references therein.

A. CMA_3/DAPI (CDD) Fluorescent sequential banding—one step

1. Dehydrate the slides through an ice-cold ethanol series 70%, 90% and 100% (2 minutes each). Let them dry out.
   Work in the dark from this step.
2. Incubate in McIlvaine buffer with $MgCL_2$ 10 minutes at room temperature. Do not let the slides dry out between step 2 and 3.
3. Stain with CMA_3 in a wet chamber for 15 minutes, at room temperature (place the slides horizontally, put 150 µl of CMA_3 solution per slide, cover with 24 x 60 mm coverslip).
4. Wash up shortly in McIlvaine buffer and allow the cover slip to detach.
5. Incubate in McIlvaine buffer with Methyl Green, 15 minutes at room temperature.
6. Wash slides shortly in McIlvaine buffer.
7. Stain with DAPI in wet chamber, for 20 minutes, at room temperature) (place the slides horizontally, put several drops of DAPI solution, cover with 24 x 60 mm cover slip).

8. Wash slides shortly in McIlvaine buffer, remove drops by touching slide margins to napkin, mount with antifade (glycerol-propylgalate) and seal the margins with nail polish. Sealed slides placed on a tray covered by aluminium foil can be kept in the fridge several days before examination.

B. CMA_3/DAPI (CDD) Fluorescent sequential banding—two steps

CMA_3-staining

1. Incubate the slides in McIlvaine buffer with $MgCl_2$ 10 minutes at room temperature. Do not let them completely dry out between step 1 and 2. Work in the dark from this step.
2. Stain with CMA_3 in wet chamber, for 15 minutes, at room temperature (place the slides horizontally; use 150 µl of CMA_3 solution per slide, cover with 24 x 60 mm cover slip).
3. Wash up shortly in McIlvaine buffer and allow the cover slip to detach.
4. Incubate in freshly prepared HEPES/NaCl buffer with Methyl Green, for 15 minutes at room temperature.
5. Wash shortly in HEPES/NaCl buffer, remove drops by touching slide margins with soft tissue, mount in glycerol with antifade (glycerol-propylgalate) and seal the margins with nail polish. Sealed slides placed on a tray covered by aluminium foil can be kept in the fridge for several days before examination.

Destaining of slides after examination

1. Remove the coverslip mechanically using a fine (ophthalmological) lancet.
2. Wash slide in xylene 2 minutes and then let them dry out shortly.
3. Soon after that, wash in distilled water, let it dry out.
4. Wash in petroleum benzine 2 minutes.
5. Wash in distilled water, 3 times.
6. Let the slides dry out for at least 30 minutes.
7. Place the slides in 70% ethanol for 3 minutes.
8. Shortly wash in distilled water.
9. Let the slides dry out and store until next use.

DAPI-staining

1. Incubate the slides in McIlvain buffer with $MgCl_2$ for 10 minutes, at room temperature. Do not let slides completely dry out between step 1 and step 2.
   Work in the dark from this step.
2. Incubate in McIlvaine buffer with Methyl Green for 15 minutes at room temperature.
3. Wash shortly in McIlvaine buffer, do not let it dry out.

4. Stain with DAPI in wet chamber for 20 minutes, at room temperature (place the slides horizontally, put several drops of DAPI solution per slide, cover with 24 x 60 mm coverslip).
5. Wash shortly in McIlvaine buffer, do not let the slides dry out; remove drops by touching slide margins with soft tissue, mount in antifade (glycerol-propylgalate), seal the margins with nail polish. Sealed slides placed into a tray covered by aluminium foil can be kept in the fridge several days before examination.
6. Destain as described above.
   Stain the slides with Giemsa according to protocol in Chapter 3.

Destaining of Giemsa-stained slides

1. Remove the immersion oil by washing the slides in xylene for 2 minutes, let it dry out, then quickly wash in distilled water, let it dry out, wash in petroleum benzine 2 minutes, let them dry out.
2. Place in 3:1 methanol:acetic acid fixative for 3 minutes, wash in distilled water, let the slides dry out and store for next use.

C-banding

This protocol follows Haaf and Schmid (1984). Alternative protocols for C-banding at Chapter 13.

1. Incubate the slides in 0.2 N HCl for 18 minutes at room temperature.
2. Wash in distilled water and let them dry out for 15 to 20 minutes.
3. Incubate in fresh prepared solution of $Ba(OH)_2$, at 33°C, for 4.5 minutes.
4. Wash the slides in 0.2 N HCl.
5. Wash in distilled water.
6. Incubate the slides in 2x SSC warmed up for 2 hours.
7. Wash in distilled water.
8. Stain 10 minutes in Giemsa, according to standard protocol.

Ag-NOR staining

This protocol follows Howell and Black (1980).

1. Adjust a hot plate to 45°C–50°C, cover surface with filter paper.
2. Place each slide (either unstained or Giemsa stained) on the paper and leave them to warm up.
3. Carefully drop two to three drops of gelatin solution on the slide by a Pasteur pipette avoiding creation of bubbles.
4. Carefully drop four to six drops of freshly prepared 50% $AgNO_3$ solution onto the same slide with a Pasteur pipette.

5. Cover them with 24 x 60 mm coverslip and carefully check the change in colour of the mixture onto the slide until it reaches a yellow-brown nuance (usually it is fast process within some tens of seconds).
6. Remove the cover slip by gently rinsing with distilled water, wash again in distilled water and let them dry out before examination.

Destaining of Ag-NOR marks

1. Place slide in A:B de-staining solutions mixture (1:9), for 4–5 minutes.
2. Wash in distilled water.
   Examples of results are shown in Fig. 1.

Troubleshooting

A number of problems may be encountered when applying sequential banding procedures. They are basically related to the use of fluorescent dyes, requiring to work in the dark. Moreover some dyes fade very quickly and the suitable antifades have to be tested.

During the observation with the microscope, attention has to be payed in order to not allow mixing of the immersion oil with the mounting medium (when present).

Prepared slides can be stored at +4°C for several days but they must be left in the dark at room temperature to adjust temperature before microscopic examination and application of immersion oil. DAPI banding can be performed as given in our protocol or with DAPI mixed in the mounting medium (e.g., Vectashield with DAPI). According to our experiences, however, the latter produces swollen, damaged chromosomes that are unsuitable for further bandings. For CBG the following points are critical. The denaturation time given in this protocol (4.5 minutes) should be adjusted according to the following: if, at the end of the procedure the chromosomes look like normal Giemsa-stained chromosomes (dark, good morphology), $BaOH_2$-treatment should be prolonged; if, instead, they result pale and swollen, this treatment should be shortened. $Ba(OH)_2$ must be always filtered; if the solution produces a film at the surface, it must be removed by paper tissue to prevent adhesion to the slides. Application of $AgNO_3$ frequently leads to precipitation of silver grains due to overheating and/or presence of dirtiness on the glass surface. After Ag-NOR staining this problem can be partially solved by short (3–4 minutes) wash of the slide in 5% $Na_2S_2O_3$ solution after examination of the staining quality.

The above problems are common to protocols either for single or sequential banding. However some of them apply specifically to sequential banding, such as removing the immersion oil applied directly to the slide, removing coverslips and mounting medium. Destaining the slides is also a critical step. The manipulation of slides in all steps requires adequate pratice and good handling in order to avoid damaging the samples. Washing and destaining chemicals (xylene, petroleum benzine, ethanol, distilled water) must be always new, in order to avoid contamination by immersion oil, mounting medium and/or dyes. Sometimes even freshly prepared destaining/washing chemicals can cause problems; if so new batches must be tested.

Laboratory set up

Equipment

Major

Fridge (4°C); freezer; water bath, adjustable to 65°C; hot plate, adjustable to 50°C.

Minor

Cuvettes (plastic or porcelain); beakers; graduated cylinders; laboratory bottles with caps; wet chamber (a metal box with cover can be appropriate to stain in the dark); fine curved forceps; ophthalmologic fine lancet; pipettes; micropipette (10–100 μl) with tips; filter paper; 24 x 60 mm cover slips; timer; marker/pencil.

Biochemicals and reagents

70%, 90%, 100% ethanol; McIlvaine buffer; McIlvaine buffer with $MgCl_2$; HEPES/NaCl buffer; CMA_3; DAPI; methyl green; xylene; petroleum benzene (medical); glycerol with propylgalate (mounting medium with antifade); nail polish; methanol; acetic acid; 2x SSC; $BA(OH)_2$; gelatin; $AgNO_3$; $AgNO_3$ destaining solution A; $AgNO_3$ destaining solution B; 0.2 N HCl.

Stock and working solutions

Methyl Green solution for CMA_3

12 g Methyl Green/100 ml HEPES/NaCl buffer.

CMA_3 solution

5 mg CMA_3/10 ml McIlvaine's buffer with $MgCl_2$
Make 0.5 ml aliquots and store frozen in Eppendorf tubes, one aliquot is for three slides.

DAPI solution

Stock solution 1 mg DAPI/10 ml McIlvaine's buffer
make 0.5 ml aliquots and store frozen in Eppendorf tubes
Working solution 5 μl DAPI stock solution/1 ml McIlvaine's buffer.

Methyl Green solution for DAPI

35 mg/100 ml McIlvaine's buffer
stock solution: 1 N HCl 44 ml HCl/500 ml D_2O
working solution 0.2 N HCl: 20 ml 1 N HCl/100 ml D_2O.

$Ba(OH)_2$ solution

5.3 g $Ba(OH)_2$/100 ml
Do not forget to filter!

Fresh aqueous 50% $AgNO_3$

0.25 g $AgNO_3$/0.5 ml D_2O is amount for 3 slides

$AgNO_3$ Destaining solutions

Solution A: 0.5% $Na_2S_2O_3$ in D_2O
Solution B: 0.5% K_3 [Fe $(CN)_6$] in D_2O
working solution 1A:9B

Mounting medium

50 ml glycerol + 1.25 g propylgalate
Dissolve in 60°C water bath overnight, store in the dark at room temperature.

References

Blaxhall, P.C. 1983. Chromosome karyotyping of fish using conventional and G-banding methods. J. Fish. Biol. 22: 417–424.

Fujiwara, A., S. Abe, E. Yamaha, F. Yamazaki and M.C. Yoshida. 1998. Chromosomal localization and heterochromatin association of ribosomal RNA gene loci and silver-stained nucleolar organizer regions in salmonid fishes. Chromosome Res. 6: 463–471.

Gold, J.R., C.Z. Li, N.S. Shipley and P.K. Powers. 1990. Improved methods for working with fish chromomes with a review of metaphase chromosome banding. J. Fish Biol. 37: 563–575.

Gromicho, M., C. Ozouf-Costaz and M.J. Collares-Pereira. 2005. Lack of correspondence between CMA -, Ag-positive signals and 28S rDNA loci in two Iberian minnows (Teleostei, Cyprinidae) evidenced by sequential banding. Cytogenet. Genome Res. 109: 507–511.

Haaf, T. and M. Schmid. 1984. An early stage of ZW/ZZ sex chromosome differentiation in *Poecilia sphenops* var. *melanistica* (Poeciliidae, Cyprinodontiformes). Chromosoma. 89: 37–41.

Howell, W.M. and D.A. Black. 1980. Controlled silver-staining of nucleolus organizer regions with a protective colloidal developer: A 1-step method. Experientia. 36: 1014–1015.

Martins, C. and P.M. Galetti. 1999. Chromosomal localization of 5S rDNA genes in Leporinus fish (Anostomidae, Characiformes). Chromosome Res. 7: 363–367.

Mayr, B., P. Ráb and M. Kalat. 1985. Localization of NORs and counterstain-enhanced fluorescence studies in *Perca fluviatilis* (Pisces, Percidae). Genetica. 67: 51–56.

Mayr, B., P. Ráb and M. Kalat. 1986. Localisation of NORs and counterstain-enhanced fluorescence studies in *Salmo gairdneri* and *Salmo trutta* (Pisces, Salmonidae). Theor. Appl. Genet. 71: 703–707.

Medrano, L., G. Bernardi, J. Couturier, B. Dutrillaux and G. Bernardi. 1988. Chromosome banding and genome compartmentalization in fishes. Chromosoma. 96: 178–183.

Rábová, M., P. Ráb and C. Ozouf-Costaz. 2001. Extensive polymorphism and chromosomal characteristics of ribosomal DNA in a loach fish, *Cobitis vardarensis* (Ostariophysi, Cobitidae) detected by different banding techniques and fluorescence *in situ* hybridization (FISH). Genetica. 111(1-3): 413–422.

Schmid, M. and M. Guttenbach. 1988. Evolutionary diversity of reverse fluorescent bands in vertebrates. Chromosoma. 97: 101–114.

Schmid, M., C. Steinlein, I. Nanda and J.T. Epplen. 1990. Chromosome banding in amphibia. pp. 2–45. *In*: Olmo, E. (ed.). Cytogenetics of Amphibians and Reptiles. Birkhäuser, Basel.

Schweizer, D. 1980. Simultaneous fluorescent staining of R bands and specific heterochromatic regions (DA-DAPI bands) in human chromosomes. Cytogenet. Cell Genet. 27: 190–193.

Schweizer, D. 1981. Counterstain-enhanced chromosome banding. Human Genet. 57: 1–14.

Seabright, M. 1971. A rapid banding technique for human chromosomes. Lancet 2: 971–972.

Shaffer, L.G. and J.R. Lupski. 2000. Molecular mechanisms for constitutional chromosomal rearrangements in humans. Annual Rev. Genet. 34: 297–329.

Sola, L., A.R. Rossi, V. Iaselli, E.M. Rasch and J.P. Monaco. 1992. Cytogenetics of bisexual/unisexual species of *Poecilia*. II. Analysis of heterochromatin and nucleolar organizer regions in *Poecilia mexicana mexicana* by C-banding and DAPI, quinacrine, chromomycin A3, and silver staining. Cytogenet. Cell. Genet. 60: 229–235.

Sumner, A.T. 1990. Chromosome Banding. Unwin Hyman, London.

Sumner, A.T., H.J. Evans and R.A. Buckland. 1971. New technique for distinguishing between human chromosomes. Nature New Biol. 232: 31–32.

Yu, Q.-X., L.C. Fan, J.-X. Cui, X.-H. Ren, K. Li and X.-J. Yu. 1994. High resolution G-banding and idiogram on pachytene bivalents of rice field eels (Monopterus albus Zuiew). Science in China (Ser. B) 37(9): 1090–1096.

15

Standard Fluorescence *in situ* Hybridization Procedures

C. Bonillo,[1] *J.-P. Coutanceau,*[2,a] *H. D'Cotta,*[3] *L. Ghigliotti,*[4,*] *C. Ozouf-Costaz*[2,b] and *E. Pisano*[5]

[1]Service de Systématique Moléculaire, MNHN, CP 26, 57 rue Cuvier, 75005 Paris, France.
e-mail: bonillo@mnhn.fr

[2]CNRS, UMR 7138, IBPS, UPMC, case 5, 7 quai St Bernard, 75952 Paris Cedex 05, France.
[a]e-mail: coutance@mnhn.fr
[b]e-mail: ozouf@mnhn.fr

[3]CIRAD, UMR 116, ISEM, Campus International de Baillarguet, TA B116/A, 34398 Montpellier cedex 5, France.
e-mail: dcotta@cirad.fr

[4]Institute of Marine Sciences (ISMAR), CNR, Via De Marini 6, 16149 Genoa, Italy.
e-mail: laura.ghigliotti@gmail.com

[5]DISTAV—Dipartimento di Scienze della Terra, dell'Ambiente e della Vita, University of Genova, Viale Benedetto XV, 5 16132 Genova, Italy.
e-mail: pisano@unige.it

*Corresponding author

Principles

Fluorescence *in situ* hybridization (FISH) allows nucleic acid sequences to be visualized inside fixed cells or chromosomes. The method, based on the principle that nucleic acids can denature and renature, allows the detection of a known nucleotide sequence (probe), in a cell/chromosome, thanks to its ability to hybridize the complementary nucleic-acid sequence (target) preserved *in situ*. FISH probes specific for fish are not commercially available and must be homemade. Before the hybridization experiment, the DNA probe has to be directly or indirectly labelled with a fluorochrome, so that the hybridization product can be revealed and visualized as a fluorescent signal. In direct labelling, the fluorochrome is directly bound to the DNA probe. In indirect labelling, a hapten (most frequently biotin or digoxigenin) is incorporated in the probe and immunochemically detected by a fluorophore-tagged antibody. The labelled probe and the target sequence are denatured and then combined, so that their complementary DNA sequences are annealed. When the probe and target DNA belong to the same species, the FISH is referred as *homologous*, whereas it is *heterologous* when they are derived from different species.

Applications and Limitations

FISH in fish has been developed continuously during the past decade and is now very widely used for *in situ* localization, with a great variety of applications such as comparative cytogenomics for evolutionary studies; structural genomics with chromosomal localization of single copy genes and various types of repeated sequences; genome mapping in model teleost fish species, where anchoring of large insert bacterial artificial chromosome (BAC) libraries allows assignation and orientation of linkage groups to specific chromosomes (see reviews and examples in Phillips 2007; Negrisolo et al. 2008; Cioffi et al. 2012; Guyon et al. 2012).

The basic techniques of standard FISH, as described in handbooks, application guides and supplier manuals for animal and clinical cytogenetics are applicable to fish chromosomes, but require some adjustments. Fish, in the broad sense of the word, include chondrichthyans and teleosts which do not represent a monophyletic group. Teleosts in particular, include around 27000 species and are highly diversified, with a very wide range of chromosome size, numbers and formulae. Fish chromosomes are often numerous, very small and fragile. They can be damaged during fixation pre-treatments and denaturation. For these reasons, it is better to denature the probe and chromosomes separately, and to adjust the time and temperature of denaturation according to the species group under study. In this chapter, we cannot provide an overview of all procedures developed in fish cytogenetic laboratories around the world. We present here a few examples of core protocols we currently use, that have been set up according to the following criterions: rapid, inexpensive, reproducible, lower toxicity for users. Tips and tricks are provided, as for instance, how to de-hybridize a chromosomal preparation for a second FISH, before being able to establish a routine protocol.

Protocols

1. Repeated sequences and multi-copy genes

1.1. Probe preparation

Probes to detect simple sequence repeats, satellite DNAs, centromeric and telomeric repeats, transposable elements, protein coding genes present in clusters or multiple copies such as ribosomal genes, can be homemade by PCR from genomic DNA of the species under study, using specific or degenerated primers. The PCR products must be controlled by running a small quantity in a 2% agarose gel, producing a smear corresponding to the expected size of the sequence to be hybridized.

PCR probes can be labelled along with the PCR reaction, but produce small DNA quantities. In order to obtain higher amounts and to check probe quality, we recommend to clone and sequence probes prior to hybridization. PCR and cloning protocols will not be provided herein.

Before labelling, PCR probes are purified using QIAquick PCR purification kit (Qiagen) according to the supplier's protocol. Cloned probes are purified using NucleoSpin Plasmid EasyPure (Macherey-Nagel, GmbH and Co).

1.2. Probe concentration before labelling

Prior to labelling, probe DNA is generally quantified by electrophoresis by running 2 µl, on a 2% agarose gel against a control DNA (50 ng/µl), or by using a sensitive fluorometer that specifically measures DNA. Probes can be also synthesized and then directly fluorescence- or hapten-labelled, by specialized suppliers.

1.3. Probe labelling

Repeat probes and multicopy genes are generally cloned in small vector-like plasmids, sequenced and directly or indirectly labelled by nick translation.

1.3.1. Indirect labelling with biotin or digoxigenin by nick translation

We recommend the use of nick-translation kits (e.g., from Roche Diagnostics or Invitrogen) and to follow the supplier's instructions. After the nick translation reaction, the fragmentation of the labelled DNA should be controlled on agarose gel and should produce a smear with an average fragment size of 300 to 700 bp. Smaller fragments may produce a background on the slides after FISH. An additional purification step may be useful to eliminate dNTPs that have not been incorporated (§ 1.1.).

Labelled probes should be stored at –20°C or ethanol precipitated:

Ethanol precipitation

- Precipitation of the probes prior to mixing it into the hybridization buffer is recommended because they may contain rests of proteins and salts. Precipitate DNA with 2 v/v ice-cold 100% ethanol and 0.1 v/v sodium acetate 3 M, pH 5.2.
- Incubate at –20°C for 60 min. Spin for 15 min at 15,000 rpm and discard supernatant. Wash the pellet with 100% ethanol, spin 15 min at 15,000 rpm for 5 min, remove supernatant and air dry.

1.3.2. Direct labelling

Cloned probes can also be directly labelled with systems in which the dNTPs are tagged with fluorophores, for instance the Spectrum fluorophores from Vysis FISH technology (Abbott Molecular) (Storing or precipitation as in Section 1.3.1.).

1.4. Preparation of the probe hybridization mixture

Dissolve the labelled probe pellet in a concentration of 10 to 20 ng/µl in the probe hybridization buffer 1 in a water-bath at 37°C overnight, mix well before use with the tip of a micropipette.

1.5. Chromosome preparations

Chromosomes must be prepared according to standard procedures (Chapter 3 to 11). In order to optimize further FISH experiments, it is better to have the chromosomes in the central area of the slide. We therefore recommend putting a single drop of cell suspension in the very middle of the slide.

1.6. Slide pre-treatment

We recommend avoiding this step, which is usually required to reduce the background, because it may damage the fish chromosomes. However, on chromosome preparations that have not been sufficiently hypotonized or fixed, it may be useful to apply a slight pepsin pre-treatment, to remove proteins that can hinder the probe access to its target.

- Incubate slides in 2x SSC at RT for 2 min (Coplin jar).
- Incubate slides in the pepsin working solution, prepared just before use and pre-warmed at 37°C, between 2 to 5 min.
- Rinse in 2x SSC, twice for 2 min each.
- Dehydrate in ethanol series (70%, 90%, 100%), 2 min each and dry the slides.

1.7. Probe denaturation

Denature the necessary quantity of probe dissolved in the hybridization buffer (15 µl per slide) in a water bath or in a thermocycler at 75°C, 5 min. Stop the denaturation in crushed ice.

1.8. Chromosome denaturation

Pre-warm 40 ml of denaturation buffer in a coplin jar in a water bath at 68°C–72°C. Control the temperature inside the Coplin jar before starting.

Denature chromosomes by immersion between 10 sec and 2 min, according to the species studied.

- Stop the denaturation in ice-cold ethanol series (70%, 90%, 100%).
- Prior to hybridization, slides can stay for a few hours in the last ethanol bath if it is maintained ice-cold.
- Dry the slides just before hybridization.

1.9. Hybridization

With a micropipette, drop 15 µl of denatured probe onto the chromosomes and cover with a 22 x 22 mm plastic (polyethylene) coverslip. Slightly press with the tip of the cone to remove bubbles. Sealing the coverslip is not mandatory if the moist chamber is well closed. Ensure the slides are in perfect horizontal position, so that the probe is regularly distributed onto the chromosomes. Leave to hybridize at 37°C overnight.

1.10. Post hybridization washes and mounting slides

1.10.1. Indirectly labelled probe

- After hybridization in the moist chamber, the coverslip will be floating. Carefully remove it with forceps. Do not let slides dry.
- Wash in 2x SSC, pH 7 at 68°C–72°C (Coplin jar in a water-bath), for 5 min, sharp.
- Transfer to 1x PBD 1 to 2 min at room temperature and without drying the slides, proceed to detection:
- Apply 40 µl of FITC-avidin or rhodamine-anti-DIG under a 24 x 40 mm glass coverslip.
- Incubate 5 min at 37°C, in a humidified chamber. Tilt and shake each slide over a waste-can to let the coverslip fall.
- Rinse 3 times for 2 min in 1x PBD at RT.
- Carefully blot slides between two sheets of filter paper, do not let it dry.
- Put 20 µl DAPI/antifade per slide under a large (24 x 50 mm) coverslip and press firmly between two sheets of filter paper, using gloves, to remove bubbles and excess anti-fade.
- Keep the slides horizontally on a tray, in the dark, prior to microscope examination.

1.10.2. Directly labelled probes

- After hybridization, remove coverslip as described in § 1.10.1.
  Wash slides in 0.4x SSC, 0.3% Tween 20 (v/v) at 60°C, 2 min, sharp (Coplin jar).
- Wash in 2x SSC, 0.1% Tween 20 (v/v), 1 min, sharp.

- Dehydrate in ethanol series (70%, 90%, 100%), 2 min each and dry.
- Drop 20 µl of DAPI/antifade, cover with 24 x 50 mm coverslips and firmly blot between two sheets of filter paper, using gloves.
- Keep the slides horizontally on a tray, in the dark, prior to microscope examination.

Examples of FISH with some repetitive probes are shown in Fig. 1.

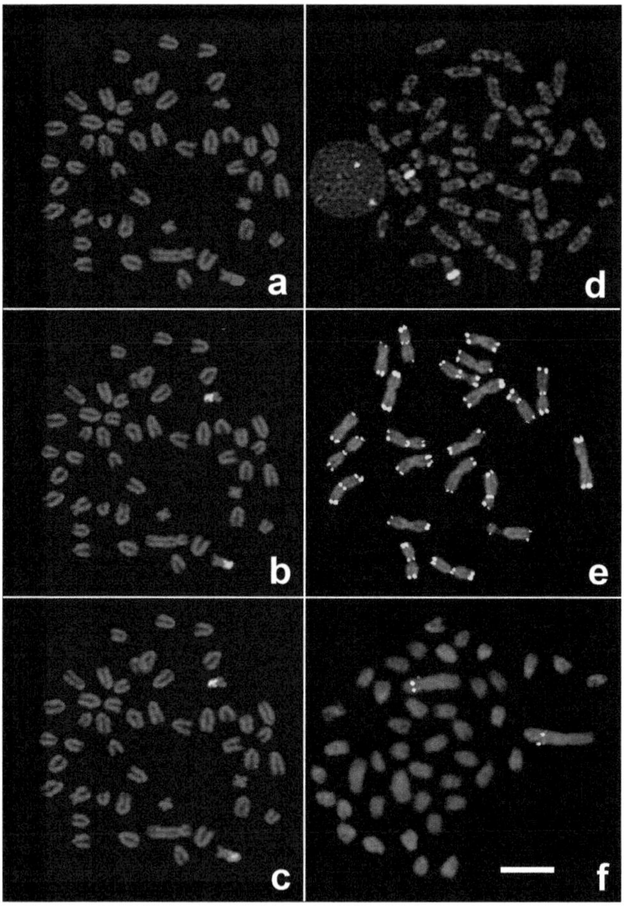

Figure 1. Examples of FISH obtained using probes indirectly labelled with biotin, detected with avidin-fluorescein (green signals) or digoxigenin, detected with anti-Dig-rhodamin (red signals). Double color FISH of 5S- and 28S rDNAs cloned in plasmids on a metaphase plate of *Chionodraco hamatus* (Channichthyinae): 5S rDNA hybridization pattern (a), 28S rDNA hybridization pattern (b), merged image (c). FISH of an immunoglobulin gene cloned in a plasmid on a metaphase plate of *Trematomus bernacchii* (Trematominae) (d). FISH of the telomeric repeat probe $(TTAGGG)_7$ on a metaphase of *Notothenia coriiceps* (Nototheniidae) (e). Double FISH of two BAC clones from the LG3 linkage group on a metaphase plate of *Oreochromis niloticus* (Cichlidae) (f). Scale bar for all pictures: 10 µm.

Color image of this figure appears in the color plate section at the end of the book.

2. Single copy gene probes

From our own experience (Pisano et al. 2003), single copy gene probes can be successfully used in FISH, if the inserted sequence is at least 1500 bp. They can be labelled and hybridized like repetitive probes, but their concentration in the hybridization buffer needs to be increased (up to 50 ng/µl) and the stringency of the hybridization buffer decreased (use hybridization buffer 2 instead of 1). The duration of the hybridization must be at least 48 hrs. If, tiny signals are present in a very low proportion of metaphase plates, the result can be considered as positive only when two clear signals are recognizable in the two chromatids of a chromosome pair. An example is shown Fig. 1d. We do not recommend amplification of the signals by fluorescent antibodies, because it generally produces background that can mismatch with the hybridization signals. According to our experience it is better to use a high sensitivity camera and enhance the signals electronically by a FISH-imaging software whenever needed. Better signals are detected with larger fragments, hence to date BAC probes are more successfully being used in FISH for low copy gene mapping applications.

3. BAC probes

Bacterial Artificial Chromosomes (BACs) are small size vectors with ability to accommodate large DNA inserts (size range 30 to 300 kb) developed to be cloned and maintained in *E. coli*. Due to their stability, efficiency of transformation and easiness of purification, along with their capability to host long inserts, BACs have been widely used for genetics and genomics, including physical mapping through FISH. Indeed, the use of BAC clones as probes for FISH is a powerful approach to anchor genomic and linkage data to physical chromosomes, thus contributing to the identification of genomic regions and to the whole-genome sequencing process. As a side effect, given their large size, the BAC inserts often contain genomic repetitive sequences (e.g., the Alu sequence in humans). *In situ* hybridization to chromosomes with a probe which contains repetitive DNA usually results in scattered hybridization spots in all chromosome pairs and sometimes uniform staining. Pre-annealing the probe with unlabelled competitor DNA (usually the genomic DNA of the target species) and carrier DNA (usually from a species with is not closely related, for instance salmon sperm DNA for human probes) prevents the repeated sequences of the probe to bind to the target, leading to more intense and specific *in situ* hybridization signals from the unique sequences of the BAC probe (Pinkel et al. 1988).

3.1. BAC probe preparation

BAC libraries have been already developed for a number of fish species so BAC clones can easily be obtained from fish genome browsers. As an alternative, species specific BAC libraries can be constructed *ad hoc* for physical mapping purposes. Details on the methods for construction, screening and clone sequencing of a fish genome BAC library go beyond the purposes of the present volume and are not provided.

BAC-clones can be fully sequenced, or only BAC-end sequenced. Normally, they are shipped in *E. coli* clones and cultivated in LB broth or better in the richer 2YT broth, supplemented with 12.5 µg/ml chloramphenicol. The DNA can be extracted from the BACs by an alkaline lysis method such as follows:

- incubate cells in LB or 2YT broth plus chloramphenicol at 37°C overnight or until OD at 600 nm is in the range of 0.4 to 0.6.
- pellet the cells through centrifugation (6000 g, 15 min, 4°C) and resuspend them in TrisHCl 50 mM pH 8/EDTA 10 mM (mix thoroughly with a vortex). RNase A is normally added to the buffer.
- add an alkaline lysis solution (NaOH 0,2 N; 1% SDS), close the tube tightly and mix only by inversion (do not vortex since it will shear the DNA).
- the digestion is performed at RT and must not exceed 5 min, otherwise the DNA may also be digested.
- remove the cell debris through acetate salt (KOAc 3M pH 4.8–5.5) precipitation.
- centrifuge at 25,000x g for 30 min at 4°C in order to remove all the precipitate and transfer the supernatant to a fresh tube. If necessary repeat the centrifugation to obtain a clear supernatant.
- collect the nucleic acids by isopropanol precipitation and centrifugation.
- resuspend in TE pH8 (TrisHCl 10 mM pH 8/EDTA 1 mM pH 8).

A number of kits are also available for BAC purification, among them we prefer the QIAGEN Plasmid Midi Kit (cat. number 12143). We routinely verify the quality of the BAC-DNA after purification by PCR since shearing of the DNA can occur. PCR can be performed if the whole BAC or a BAC-end sequencing has been done, or if a fragment/gene of the BAC is known.

3.2. BAC Probe concentration before labelling

Prior to labeling, the amount of probe DNA must be accurately estimated, by electrophoresis on 1% agarose gel in presence of control DNA and/or by using a sensitive Qubit® fluorometer which measures only DNA and not RNA-factors, thus enhancing the accuracy of the measurement.

3.3. BAC probe labelling

3.3.1. Indirect BAC probe labelling with biotin or digoxigenin

BACs can be labelled by nick translation, but we have obtained better results with random priming, using the "High-Prime DNA labelling" kits from Roche

Diagnostics. Before labelling following the supplier's protocol, the probe must be sheared 30 min at 95°C in a thermocycler and chilled in crushed ice; this is a critical step.

3.3.2. Direct BAC probe labelling with fluorophores

It is possible to directly label BACs using Spectrum fluors, according to the protocol provided in §1. In this section we will limit descriptions to labelling with Kreatech "PLATINUM Bright labelling kits" (Leica Biosystems) which proposes five different fluorophores. These kits produce a very clear and bright labelling using BACs, with a small modification: prior to the first step provided in the supplier's protocol, BACs must be sheared using the same protocol as described in § 3.3.1.

3.4. Competitors

For teleosts, we hybridize BAC probes at around 2.5 ng/µl in the hybridization buffer after pre-annealing them with a 200-fold of specific sonicated competitor DNA and a 1000 fold carrier DNA (generally bovine).

Competitor and carrier DNA preparation:

Large quantities of DNA are extracted from fish and bovine tissue by the CTAB method (Winnepenninckx et al. 1993) and quantified (see protocol § 1.2.).

Carrier and competitor DNAs are heated at 120°C a few seconds in an autoclave, chilled in crushed ice and ethanol precipitated (see protocol § 1.3.1.) before being prepared in stock solutions at suitable concentrations (see Section "Biochemicals and Reagents" at the end of this chapter).

3.5. Probe denaturation and reannealing

Pre-warm probe, competitor and carrier DNA stock solutions at 37°C in a water-bath.

In an Eppendorf tube, mix up 2.5 µl probe, 1 µl competitor and 4 µl carrier DNA stock mixtures, add 7.5 µl hybridization buffer mix up with the tip of a pipette cone, centrifuge briefly and denature for 5 min in a 72°C ± 2°C water-bath. Re-anneal in a water-bath 37°C for 90 min to maximum 2 hrs. Chill in crushed ice.

N.B. Competitor and carrier mixture pipetting may be facilitated by heating them around 60°C.

3.6. Chromosome denaturation

Proceed as in § 1.8.

3.7. Hybridization

Proceed as in § 1.9. immediately after pre-annealing the probe, but hybridize at 37°C during 48 hrs.

3.8. Post hybridization washes and mounting slides

3.8.1. Directly labelled probes

Proceed like in § 1.9.1.

3.8.2. Indirectly labelled probes

- After hybridization in the moist chamber, the coverslip will be floating. Carefully remove it with forceps. Do not let slides dry.
- Wash slides in 0.4x SSC, 0.3% Tween 20 (v/v) at 60°C, 2 min, sharp.
- Wash in 2x SSC, 0.1% Tween 20 (v/v), 1 min, sharp.
- Do not dry the slides and apply 40 µl of FITC-Avidin or anti-DIG-Rhodamine under 24 x 40 mm coverslip. Incubate 5 min at 37°C, humidified chamber.
- Remove coverslip as explained in § 1.9.2. (Detection).
- Rinse 3 times for 3 min each time with PBD. Dehydrate in ethanol series (70%, 90%, 100%), dry the slide and mount in DAPI/antifade under 24 x 50 mm coverslip.

4. Double and triple FISH

4.1. Double BAC-FISH

- Use the same stock solutions of BAC probe, competitor and carrier, as for simple FISH.
- In an Eppendorf tube A, gently mix up 2.5 µl probe stock solution A; 1 µl specific competitor DNA; 4 µl carrier DNA; briefly centrifuge and proceed to probe denaturation step and re-annealing.
- In an Eppendorf tube B, gently mix 2.5 µl of probe stock solution B; 1 µl specific competitor DNA; 4 µl carrier DNA; briefly centrifuge and proceed to probe denaturation step and re-annealing.
- After re-annealing, pool A + B mixtures (carefully mix together with the pipette tip) and proceed to hybridization and post-hybridization washings in the same conditions as with a single probe.
- If the two probes have been indirectly labelled, detect signals with 40 µl of Avidin-FITC/Rhodamin anti DIG. Next steps are the same as for a single probe. An example of double FISH by using Bac probes is shown in Fig. 1f.

4.2. Triple BAC-FISH

- Each probe must be directly labelled and its concentration in the hybridization buffer must be the same as for simple fish.
- Fluorophore excitation-emission wavelengths must not be overlapping.

Troubleshooting

Slide preparation

We recommend for beginners to test these techniques with chromosome preparations that are easily available, and to check the amount of well spread metaphase plates in the preparation by microscope observation in phase contrast at magnification x10 or x20.

Slide pre-treatment with pepsin solution

Pepsin pre-treatment, except if rests of cytoplasm are clearly visible, must be avoided. Time of pre-treatment with this protease is very variable according to fixation, age of the slide, etc. If several slides are to be treated, pre-warm several Coplin jars in a water bath and use a different jar for each slide.

Nick-translation

It is better to re-suspend the probe pellet in bidistilled water instead of TE, because EDTA can hinder nick-translation reaction.

Probe denaturation

As a general rule in teleosts, the smaller the chromosomes, the shorter the denaturation. Over denaturation is a frequent problem with teleost fish chromosomes. In this case, chromosomes appear puffy, surrounded by a cloud of DNA filaments, thus making it impossible to precisely localize the FISH signals in a specific chromosomal region.

Hybridization

As 22 x 22 mm coverslips, we use polyethylene plastic foils, thickness 50 μm, cut out with scissors. Glass ones can also be used, but they are more difficult to remove after the hybridization step.

Most repetitive probes (ribosomal, telomeric…) hybridize within two hours. Leaving them overnight is just a precaution. Leaving it more than 24 h may result in cross-hybridization and excess of background.

For single copy gene probes, duration of hybridization can be increased to three or four days with a higher probability for the probe to bind to its target without a cross hybridization problem.

BACs generally produce clear FISH signals within 48 h, but increasing this time may produce background since they often contain repetitive sequences, even after re-annealing the probes.

It is not necessary to seal the coverslips on the slides: if the moist chamber is maintained horizontally and well closed, with a sufficient quantity of distilled water in its lower part. Replacing this distilled water by formamide and/or 2x SSC is useless and pretty obsolete.

FITC-Avidin/anti-DIG Rhodamine

Optimum pH and working concentrations are different for anti DIG-rhodamine and FITC-avidin. For double FISH, an intermediate pH is used, but each of the two fluorochromes must be used at their own working final concentration. These two fluorochromes may be filtered before use with 0.45 µm single use syringe filters to avoid occurrence of crystal deposits on the chromosome preparations.

Absence of FISH signals

They can be due to insufficient chromosome denaturation, but also to too stringent, to hot or too long washes. Probe concentration must be controlled and possibly increased. It can also be due to a failure in probe labelling or to a defective labelling kit.

Excess of FISH signals

This problem frequently occurs with some BACs containing a high amount of repetitive sequences, even in the presence of competitor and career DNA. It can be verified by entirely sequencing the BAC (generally, only the BAC ends are sequenced). In this case, it is better to use another BAC matching for the same chromosome region.

DAPI-antifade counterstaining

The commercial DAPI-Antifade solutions are made for mammal chromosomes and not sufficiently concentrated for fish chromosomes. We recommend preparing a small quantity, to test the staining on a fish chromosome preparation and to eventually increase the DAPI concentration so that metaphase plates could be easily recognized at microscope magnification x20.

Successive hybridizations on a same chromosome preparation

This can be useful if a FISH experiment must be repeated with protocol modifications or if several probes of different types need to be hybridized: wash the hybridized slide in PBD until the coverslip falls; wash 5 min in a new PBD bath, slightly shake at the end of the washing; wash 3 min in 4x SSC, at RT; wash 3 min in 2x SSC, at RT; dehydrate in 3 ethanol bathes, 70%, 80%, 100%, 2 min each. Leave slides overnight in the last ethanol bath at +4°C. Dry and store in deep freezer or denaturate and re-hybridize another probe.

Storing slide preparations after FISH experiment

We recommend not leaving FISH slides at room temperature for more that one week, if they cannot be rapidly examined. For long-time storing, coverslips must be sealed with nail polish and cleaned with ethanol if they have been covered with immersion oil. Slides can be stored at –20°C in the dark for further microscope examinations.

Laboratory set up

Equipment

Major

Thermocycler; ultra-centrifuge; Gel electrophoresis equipment (Mupid One); sensitive fluorometer; deep freezer; water bath; refrigerator; incubator; crushed ice machine; fume hood; fluorescence microscope equipped with high sensitivity camera and objectives x20 and x100.

Minor

Eppendorf tubes; micropipettes and cones; vortex; slides; coverslips; coplin jars; water bath thermometers; filter paper; moist chamber; flasks and cylinders; syringe filters "minisart RC15", 0.45 µm, Sartorius, Cat. 17762.

Biochemicals and Reagents

Ethanol; sodium acetate; Tween 20; Anti-DIG-rhodamine, Fab fragments (Cat. 1 207 750, Roche, sold by 200 µg frozen, lyophylized); streptavidin fluorescein (Cat. No. 1 975 595, Roche, sold by 1 mg frozen, lyophylized); Antifade Vectashield mounting medium (cat. H-100, Vector Labs); DAPI, (SIGMA cat. D9542); Agarose, DNA-grade; LB broth with chloramphenicol (Teknova cat. L8076); 2YT broth with chloramphenicol-34, sterile (Teknova cat. Y1175); Tris/HCl; EDTA; Sodium dodecyl sulfate (SDS); RNase A; NaOH; potassium acetate salt (CH3COONa); isopropanol.

Stock and working solutions

2x SSC

PBD

For 100 ml: 1X PBD, mix 20 ml 20x SSC; 1 ml Tween 20, distilled water qsp 100 ml; adjust to pH 7. Store a few days at +4°C. Remember to shake the flask before each use.

Probe hybridization buffer 1

Dissolve 4 g of dextran sulphate in 20 ml 70% dionized formamide/2x SSC/50 mM phosphate buffer in a waterbath at 60°C with frequent vortexings during 3 to 5 hrs. Aliquot in 2 ml Eppendorf tubes and store at –20°C.

Probe hybridization buffer 2

Same components as for probe hybridization buffer 1, but use 50% dionized formamide.

Pepsin stock solution: 10% pepsin in distilled water can be prepared at one time, aliquoted and stored for years at –20° C.

Pepsin working solution

0.005% enzyme in 0.01 N HCl.

Competitor DNA stock solution

Specific genomic DNA, sonicated into 300–800 bp fragments and precipitated in ethanol: prepare, as for the probe, a stock solution in probe hybridization buffer 2 at a concentration of 8 µg/µl.

Carrier DNA stock solution

Use human or bovine DNA, sonicated into 300–800 bp fragments, prepare a 10 µg/µl stock solution in probe hybridization buffer 2. You may heat the solution up to 60°C–70°C to dissolve it correctly.

Anti-DIG rhodamin

Stock solution: Dissolve the lyophylized pellet in 1 ml bidistilled water to obtain a 200 µg/ml stock solution. Aliquot per 10 µl in well closed Eppendorf 2 ml tubes. Store at –15°C/–25°C in the dark.

Working solution: Complete an aliquot at 1 ml with PBD 1X, pH 8.4. For double coulour FISH: use half of this working solution, half of the Avidin/FITC working solution (for instance, 20 µl of each for a 24 x 40 mm coverslip).

For simple FISH, use half of this working solution, half of PBD 1X pH 8.4. This solution stays stable up to 2 months at +4°C.

FITC-Avidin

Stock solution: Dissolve the lyophylized pellet in 1 ml bidistilled water to a 1 mg/ml stock solution. Aliquot per 20 µl in well closed Eppendorf 2 ml tubes. Store at –15°C/–25°C in the dark.

Working solution: Complete an aliquot at 1 ml with PBD 1X, pH 7. For double colour FISH: use half of this working solution, half of the Rhodamin anti-DIG working solution (for instance, 20 µl of each for a 24 x 40 mm coverslip). For simple FISH: use half of this working solution, half of PBD, pH 7. This solution stays stable up to 2 months at +4°C.

DAPI/antifade

In an Eppendorf tube, mix up by vortexing: 1ml of antifade, 12 µl of a 100 µg/ml DAPI solution in 2X SSC (stored at –20°C in aliquots, in the dark). Store the DAPI/antifade solution in the dark at +4°C.

References

Cioffi, M., W.F. Molina, R.F. Artoni and L.A.C. Bertollo. 2012. Chromosomes as tools for discovering biodiversity—the case of Erythrinidae Fish Family. *In*: Tirunilai, P. (ed.). Recent Trends in Cytogenetic Studies—Methodologies and Applications, InTech, Available from: http:// www.intechopen.

Guyon, R., M. Rakotomanga, N. Azzouzi, J.P. Coutanceau, C. Bonillo, H. D'Cotta, E. Pepey, L. Soler, M. Rodier-Goud, H. D'Hont, M.A. Conte, N.E.M. Van Bers, D.J. Penman, C. Hitte, M.A. Crooijmans, T.D. Kocher, C. Ozouf-Costaz, J.F. Baroiller and F. Galibert. 2012. A high-resolution map of the tilapia genome: a resource for studying cichlids and other percomorphs. BMC Genomics 13: 222.

Negrisolo, E., L., Bargelloni, T. Patarnello, C. Ozouf-Costaz, E. Pisano, G. DiPrisco and C. Verde. 2008. Comparative and evolutionary genomics of globin genes in fish. Methods in Enzymology 436: 507–534.

Phillips, R. 2007. Application of fluorescence *in situ* hybridization (FISH) to genome mapping in Fishes. pp. 455–472. *In*: Pisano, E., C. Ozouf-Costaz, F. Foresti and B.G. Kapoor (eds.). Fish Cytogenetics. Science Publishers, Enfield, USA.

Pinkel, D., J. Landegent, C. Collins, J. Fuscoe, R. Segraves, J. Lucas and J. Gray. 1988. Fluorescence in situ hybridization with human chromosome-specific libraries: detection of trisomy 21 and translocations of chromosome 4. Proc. Nat. Acad. Sci. 85(23): 9138–9142.

Pisano, E., E. Cocca, F. Mazzei, L. Ghigliotti, G. Di Prisco, H.W. Detrich III and C. Ozouf-Costaz. 2003. Mapping of α- and β-globin genes on Antarctic fish chromosomes by fluorescence *in situ* hybridization. Chrom. Res. 11: 633–640.

Winnepenninckx, B., T. Backeljau and R.D. Wachter. 1993. Extraction of high molecular weight DNA from molluscs. T I G 9: 407.

16

Characterization of Fish Genomes by GISH and CGH

R. Symonová,[1,#,*] *A. Sember,*[1,2] *Z. Majtánová*[1,3] and *P. Ráb*[1]

[1]Laboratory of Fish Genetics, Institute of Animal Physiology and Genetics, Academy of Sciences of the Czech Republic, 277 21 Liběchov, Czech Republic.

[2]Department of Genetics and Microbiology, Faculty of Science, Charles University in Prague, Viničná 5, 128 44 Prague 2, Czech Republic.

[3]Department of Zoology, Faculty of Science, Charles University in Prague, Viničná 7, 128 44 Prague 2, Czech Republic.

[#]Current address: Research Institute for Limnology, University of Innsbruck, Mondseestrasse 9, A-5310 Mondsee, Austria

*Corresponding author: Radka.Symonova@uibk.ac.at

Principles

Using total genomic DNA (gDNA) as a probe for *in situ* hybridization (ISH) experiments is a modified FISH (fluorescence *in situ* hybridization) technique that becomes an increasingly relevant approach in non-clinical research, including fish molecular cytogenetics. GISH (genomic *in situ* hybridization) and CGH (comparative genomic hybridization) are two basic types of ISH experiments using gDNA as probes. These methods utilize DNA-DNA hybridization and sequence similarities to visualize genome specific DNA on metaphase spreads. They allow for comprehensive analysis of entire genomes:

1. In GISH the labelled gDNA of one organism is used as a probe and hybridized to target chromosomal DNA of another, usually a hybrid organism, to discriminate chromosomes originating in one of the parental species.
2. CGH enables the detection of unbalanced chromosomal rearrangements. In clinical application, one reference gDNA probe (healthy control) and one studied gDNA probe (patient, DNA from tumor cells) are cohybridized to reference metaphase spreads only. In molecular cytogenetics this method enables comparison of two differentially labelled gDNA probes on

chromosomes of one or both tested genomes (i.e., reciprocal scheme when chromosomes preparations of both genomes are available). In principle, in a typical CGH experiment, one of the compared gDNA always belongs to the genome that is identical with the genome of chromosomes tested. The experiment may be designed as intraspecies or interspecies CGH. The intraspecies design may involve a comparison of male and female gDNA to identify sex-specific chromosomes (regions) or a comparison of a tumour genome (patient) with healthy DNA (reference) on normal chromosomes. The interspecies design may involve gDNA of two or several species compared against each other. For an example and specific details in fish see Symonová et al. (2013a).

A variety of modifications of CGH and GISH exist based on the tested system and DNA/chromosome probes available, in the end blurring the differences between these two approaches. Actually, the most striking difference is in the target genome—a hybrid of compared organisms (mostly GISH) or a pure species more or less related to the compared organisms (mostly CGH). Both approaches employ an unlabelled competitor DNA, *sensu* Sealey et al. (1985) or the Cot1-DNA, *sensu* Landegent et al. (1987). Total gDNA of another species, phylogenetically not closely related, is used unlabelled and sheared (physically or enzymatically) to about 100 bp long fragments as competitor DNA to suppress hybridization of unspecific repetitive sequences, blurring the species' specific signal.

Applications and Limitations

GISH has been routinely used especially in plant cytogenetics to discriminate parental genomes in allopolyploids, e.g., Takahashi et al. (1999), Matoba et al. (2007) and Lim et al. (2007). In conventional GISH experiments, labelled gDNA from one species and unlabelled competitor DNA from another species are simultaneously applied to chromosome spreads to map chromosomal distribution of DNA of the labelled genome. Later, comparative GISH (cGISH) technology has been introduced in plants She et al. (2012). In cGISH experiments, the labelled gDNA of one species hybridizes to chromosomes of more or less related species without the application of competitor DNA; thus, the hybridization signals represent DNA sequences in common between the two species. Therefore, cGISH technology is a useful tool for comparative genome analysis to visualize chromosomal distribution of conserved DNA sequences.

CGH has been developed as a tool for solid tumour cytogenetics, to analyze chromosomal imbalances in entire genomes (Kallioniemi et al. 1992). Since cells of solid tumours usually do not produce usable chromosome spreads, it is possible to compare their genome with a reference (healthy) genome in the form of gDNA to normal (healthy) chromosomes. An analogous situation can be found, when an entire genome comparison of more fish species is desirable.

However, in some species no usable chromosome spreads are available. In this case, only their gDNA can be used for comparison with chromosomes of other species.

In fishes, male/female CGH was applied to mitotic and meiotic chromosomes of the zebrafish *Danio rerio*, the platyfish *Xiphophorus maculatus* and the guppy *Poecilia reticulata* to localize sex-specific chromosome regions and track early stages of sex chromosome differentiation (Traut and Winking 2001). Application of CGH has also proven to be a good tool in genomic comparison within sympatric species pairs of *Coregonus* as shown by Symonová et al. (2013b). Cross-species GISH using labelled cDNA of tilapia *Oreochromis niloticus* were used to compare syntenic chromosomal regions across the genomes of several South American and African cichlid species (Valente et al. 2009). Identification of species-specific chromosomes in an allotetraploid gynogenetically reproducing female genome of *Carassius auratus gibelio* was documented using GISH by Zhu and Gui (2007). A similar study was performed by Knytl et al. (2013) in European *Carassius* and revealed a natural allotetraploid female.

Another example of the useful application of GISH in fishes is the identification of nuclear genomic constituents in hybrid spined loaches. Across Europe, the spined loach of the genus *Cobitis* produced 2n, 3n and 4n clonal lineages. The all-female clones originated by hybridization and subsequent genome addition in pre-Holocene times and reproduce by gynogenesis. The high morphological, and in some cases biochemical, similarity makes proper identification of their biotypes and studies on chromosomal rearrangements difficult. For instance, genomes of *C. tanaitica* and *C. taenia* in their hybrids cannot be distinguished by allozyme studies, but they can be clearly identified by GISH. Moreover, homologous and homoeologous chromosomes of hybrids cannot be distinguished in the normal DAPI stained metaphase. Here, as an example, we show results of a GISH experiment with labelled gDNA from both parental species as probes to discriminate genomic composition of the *C. 2elongatoides-1tanaitica* hybrid. GISH clearly distinguishes the two haploid sets of the *C. elongatoides* genome and one haploid set of the *C. tanaitica* chromosomes in this hybrid (see Fig. 1).

Protocol

The following protocols are modified combinations of procedures described by Cremer et al. (2008), Bi and Bogart (2006), Traut et al. (1999) and references therein.

Chromosome spreads should be prepared in advance, so that chromosome aging can be carried out.

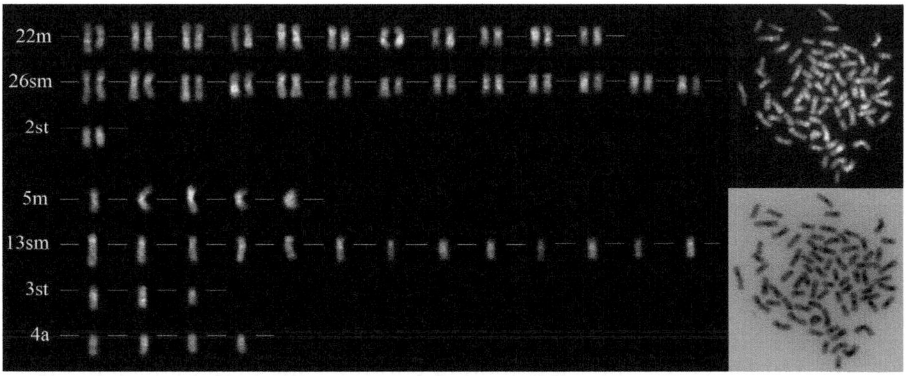

Figure 1. Representative metaphases of the *Cobitis 2elongatoides-1tanaitica* hybrid visualized by GISH with *C. elongatoides* and *C. tanaitica* genomic probes. Homologous and homoeologous chromosomes of triploid hybrids cannot be distinguished in normal DAPI stained metaphase, however, the GISH clearly distinguishes the *C. elongatoides* (shown as those with green fluorescence) and *C. tanaitica* chromosomes (shown with red fluorescence) in *C. 2elongatoides-1tanaitica* hybrid.

Color image of this figure appears in the color plate section at the end of the book.

1. gDNA isolation and specific competitor/Cot1-DNA preparation

The phenol-chloroform method of gDNA isolation, *sensu* Graham (1978) can be employed as well as using commercial kits, e.g., DNeasy Blood & Tissue Kit (Qiagen). The total DNA amount required for one GISH/CGH experiment is related to the amount of the competitor or Cot1-DNA. According to standardized protocols, e.g., Cremer et al. (2008), 1 μg of gDNA per genome probe is prepared. Therefore, when using a competitor or Cot1-DNA in 3 to 50 times higher amount than the probe, as much as 4 to 51 μg of gDNA is required for one experiment and one genome. Application of competitor DNA involves using additional unlabelled gDNA sonicated into 100–200 bp long fragments in the final concentration 3 to 50 times higher than the probe. The intensity and number of cycles of sonication has to be tested for each DNA probe individually. A crucial step is cooling the vessel containing DNA after each cycle of sonication to avoid an excess of degradation. The concentration of the competitor DNA has to be measured after the sonication, because during this step, some DNA can be lost. The preparation of the Cot1-DNA is described by Trifonov et al. (2009).

In this step, a way of gDNA amplification may become useful, e.g., GenomiPhi DNA amplification kits by GE Healthcare or Amersham Biosciences as well as a way of DNA concentration using a vacuum concentrator machine.

Particularly for CGH, it is crucial to isolate a higher amount of gDNA of very good quality in terms of its purity and similar fragment size. Therefore, the conventional phenol-chloroform method is rather suggested for this step. In the case only when a little amount of the starting material (tissue) is available, the Chelex® 100 gDNA extraction can be employed as described by Walsh et al. (1991).

2. Chromosome spreads preparation and aging

Chromosome spreads can be produced by either of the methods described in this book. Other possibilities to produce chromosome spreads in fishes are lymphocyte culture described by Fujiwara et al. (2001) or the direct preparation from cephalic kidneys described by Ráb and Roth (1988).

To improve the adhesion of chromosomes to the glass surface and to make the structure of the chromosome compact and thus enable optimal access of the labelled DNA to the target DNA, aging of the chromosomal preparations is highly recommended. The routinely applied and the easiest way is dry heat aging through one of these incubations: 7 days at RT, 3 to 8 hrs at 37°C, or 1 to 3 hrs at 60°C. The exposition of chromosome preparations to higher temperature also reduces the effects of cytoplasm remnants over chromosome spreads, when stained with DAPI.

A possible option is chemical aging, by exposing a slide with 150–200 µl of EtOH covered with a coverslip to 94°C for 2 to 20 sec (in a thermocycler). The chemical aging can be carried out when it is desirable to shorten the whole protocol. For more details, see Henegariu et al. 2001.

3. Chromosome spread pre-treatment

Pre-treatment of chromosome spreads prior to ISH allows

- decreasing the background, which results in an unspecific signal, by removing cytoplasm remnants, RNA and proteins that can bind to probe and to detection reagents.
- improving the access of DNA probes to the target DNA.

Pre-treatment is performed according to the following steps:

a) Brief equilibration in 2x SSC, 5 min, RT.
b) RNase treatment: Drop 200 µl/slide of RNase solution (20 µg/µl), cover with cover slips and incubate for 1 to 3 hrs at 37°C.
c) Wash in PBS, 5 min, RT.
d) Pepsinization: incubate slides in a coplin jar with pepsinization solution for 1 to 3 min at 37°C in a water bath.
e) Wash in 4x SSC, 5 min at RT.

4. Probe preparation, DNA labelling by Nick Translation (NT)

The best method of gDNA probe labelling is the Nick Translation (NT). This procedure can be carried out either using a commercial kit (e.g., the Roche Nick Translation Mix, Invitrogen BioNick™ DNA Labelling System or Abbott Molecular Nick Translation Kit) according to the manufacturer's instructions or according to the following protocol (the price per reaction in the commercial kits is much higher than when chemicals are purchased separately). The DNA probe can be labelled either directly with dUTP coupled with fluorochromes (e.g., Cy™3-dUTP, FITC-dUTP, Rhodamin-dUTP, Alexa Fluor® products, etc.) or indirectly with dUTP coupled with haptens (e.g., biotin and digoxigenin). When working with a directly labelled DNA probe, more steps of the protocol must be performed protected from a direct light source because the risk of fluorochrome bleaching during the manipulations is higher; on the other side, the use of an indirectly labelled gDNA probe can increase the risk of an unspecific background during hybridization signal analysis.

NT reaction:

1 µg gDNA
5 µl 10x NT Buffer
5 µl NT dNTPs-mix
5 µl 0.1 M mercaptoethanol
1 to 2.5 µl modified dUTP 1 mM (i.e., Biotin-dUPT/Digoxigenin-dUTP)
1 µl DNA polymerase I (0.1 U/µL)
1 µl DNase I (2,000 U/mL, diluted 1:10–200 with ddH_2O)
x µl ddH_2O q.s.p. 50 µl/reaction

According to the standard protocol (e.g., Cremer et al. 2008 and/or manufacturers of the commercial kits), the NT reaction involves incubation for 90 min at 15°C. However, this time strongly depends on the starting gDNA quality and the length of its fragments. Even when gDNA of the recommended fragments size (i.e., 3 to 5 kbp) is used, the time required for reaching the proper resulting fragment size (200 to 500 bp) may vary strongly and needs to be tested in each batch/species individually prior to preparing a higher amount of probes.

For this purpose, either commercial Biotin-dUTP/Digoxigenin-dUTP (e.g., Roche) are available, or there is a possibility to perform coupling of hapten succinimidyl esters (in DMSO) to aminoallyl-dUTP as described by Cremer et al. (2008). In the latter case, the cost is 100 to 300 times lower than using commercial biotin/digoxigenin-dUTPs.

Check with the 0.8% agarose gel electrophoresis, whether the length of DNA fragments is in the range of 200 to 500 bp. Probes should be kept on ice during electrophoresis. When longer fragments occur, additional time of NT incubation would be necessary (see "Troubleshooting"). As soon as the right fragment size is obtained, stop the NT reaction by following these steps (or as indicated by the NT kit manufacturer):

1. Add 2 ul 0.5 M EDTA, pH 8, and mix well.
2. Incubate 10 min at 65°C.
3. Cool on ice.

The NT product can either be stored at –20°C for up to several years or can be immediately used as follows:

a) Mix the labelled probes that will be hybridized together. When the competitor/Cot1-DNA is used, add this DNA in the concentration 3- to

50-fold the concentration of one probe DNA. The competitor DNA can be derived from one genome only, usually from the same one as the chromosome preparation originated. However, different modifications according to experimental scheme are possible.

b) Add following reagents:
   5 µg unlabelled salmon sperm* DNA for each probe
   0.1 vol. 3 M Natrium (Sodium) Acetate pH 5.2
   2.5 vol. 96% EtOH

*when working with salmonid fishes, use BSA as blocking reagent instead of salmon sperm.

c) Mix all components gently.
d) Incubate at least for 30 min at –20°C (this step can be prolonged for up to several days).
e) Centrifuge at 13,000 RPM for 15 to 20 min.
f) Discard the supernatant.
g) Add 200 ul of 70% undenatured ethanol to wash out the saline components such as sodium acetate.
h) Centrifuge at 13,000 RPM for 15 to 20 min.
i) Discard the supernatant.
j) Dry the pellet carefully with an absorbent and antistatic laboratory wipe. Let it dry for 1 to 2 minutes at 60°C or 10 to15 min at 37°C until all the liquid has evaporated. When possible, a vacuum centrifuge can be used in this step instead of 37°C/60°C incubation.
k) Re-suspend the pellet in 20 to 30 µl of the hybridization buffer and shake overnight at 37°C. In the case of direct labelling, cover the tubes on the shaker with aluminium foil to prevent the fluorochrome from bleaching.
l) On the next day, check whether the pellet has re-suspended completely.
m) After re-suspension, the hybridization probes can be used in further steps. When they are not used, they can be stored at –20°C for up to several months.

5. Denaturation of chromosome spreads and labelled DNA probe

a) Preheat 70% formamide/2x SSC at least for 30 min at 76°C in a coplin jar.
b) Denature the chromosome spreads in the coplin jar with 70% formamide/2x SSC for 3 min at 76°C.

Note: The time and temperature may vary considerably (10 s to 4 min/72 to 76°C) and must be adjusted specifically for each fish group.

c) Dehydrate slides in 70% and 80% ice-cold ethanol (in coplin jars on ice), and in 96% ethanol, at RT, 3 min each, and let it air-dry until usage (not longer than several hours).
d) Denature the DNA probe in a 1.5 ml Eppendorf tube in a water bath or a heating block at 86°C for 6 min and place immediately on ice for at least 10 min or until use.

Note: In the case of higher or too high proportion of repetitive sequences and no possibility to add competitor/Cot1-DNA, an optional step of pre-annealing can be included to the protocol by incubating the denaturated DNA probe for 90 min at 37°C in order to suppress the repetitive sequences.

This optional step is based on the higher re-association kinetics of repetitive sequences that will preferentially re-associate within the probe DNA. The negative effect of repetitive sequences is thus suppressed by their mutual re-association. However, this step is not as effective as using the unlabelled competitor/Cot1-DNA, which should be preferred in most cases.

6. Setup of hybridization

Pipette the denatured DNA mixture (20 to 30 μl) onto the denatured chromosome preparation and carefully cover with a cover slip (24 x 50 mm). Place slides in a humid chamber (e.g., metal box with grid filled distilled water) and incubate for 48 hrs to 72 hrs at 37°C. To improve the hybridization reaction, this step can be prolonged for up to 5 days.

7. Washing and detection of indirectly labelled probes*

a) Place slides in 4x SSC/0.2% Tween 20, 5 min, RT, to help removing the cover slip.
b) Wash slides in 50% formamide 2x SSC at 45°C, twice for 10 minutes, occasionally shaking them, in coplin jars.
c) Wash slides, in 1x SCC at 43 to 45°C, 3 times, for 7 min, occasionally shaking them, in coplin jars.
d) Apply a blocking treatment (500 to 800 μl 3% BSA/4x SSC/0.2% Tween 20) onto each slide and cover with a cover slip, incubate for at least 20 min at 37°C.
e) Briefly wash in 4x SSC/0.2% Tween 20 at RT.
f) In the meanwhile centrifuge the stock solution of antibodies for 3 min at 5,000 to 8,000 RPM before preparing the following mixture.
g) First antibody layer: e.g., Anti- Digoxigenin-Fluorescein + 0.5% BSA/PBS (1:20, v/v)**.
h) Pipette 150 μl of the mixture per slide, cover with 24 x 60 mm cover slip and incubate for 45 to 60 min at 37°C.
i) Wash in 4x SSC/0.2% Tween 20 at 42 to 45°C 3 times, for 7 min, occasionally shaking them.
j) Apply blocking treatment (500 to 800 μl 3%BSA/4x SSC/0.2% Tween 20) on the slide, cover by cover slip, incubate at least for 20 min at 37°C.
k) Briefly wash in 4x SSC/0.2% Tween 20.
l) Second antibody layer: e.g., Cy™-3-Streptavidin + 10% NGS/PBS (1:100, v/v).**/***

m) Pipette 150 μl of the mixture per slide, cover with 24 x 60 mm cover slip and incubate for 45–60 min at 37°C.
n) Wash in 4x SSC/0.2% Tween 20 at 42 to 45°C 3 times, for 7 min, occasionally shaking.
o) Add a drop of DAPI containing an antifade (e.g., Vectashield) on each slide, close by a cover slip (24 x 50 mm) and seal with a colourful nail polish.

*Note 1: In the case of directly labelled DNA probes (i.e., when fluorochrome is directly incorporated into the probe using, e.g., Cy3-dUTP or Fluorescein-dUTP) you can omit steps d to m.

**Note 2: The exact concentration of the antibody should be determined by the user, manufacturer's recommendations usually indicate a higher concentration than necessary.

***Note 3: In order to shorten the detection time it is possible to apply a combined solution of antibodies of both layers at once. We recommend such a simplification only if the conventional protocol works well.

Antibodies are sensitive to light therefore the steps involving fluorochromes should be performed in darkness.

8. Microscope analysis and data acquisition

For the last steps an epifluorescent microscope equipped with filters compatible with the fluorochromes used is required. An acquisition software enabling acquisition of each of the RGB channels separately in black and white regime and enabling their superimposition in pseudo-coloured composite images in RGB is also necessary. The final digital processing can be performed by using for, e.g., Adobe Photoshop software. Details on microscopy and imaging systems for ISH methods are provided by Iourov (2009), for details on fluorescent optical filters, see Sommerauer and Feuerbacher (2009).

9. Re-using chromosome preparations

After a CGH/GISH experiment it is possible to re-use slides, e.g., for a FISH experiment or for other kinds of chromosomes staining/banding:

a) Pre-heat two coplin jars with 4x SSC/0.1% Tween at 37°C.
b) Remove the nail polish from the cover slip and slide using a lancet.
c) Incubate the slide with cover slip in the pre-heated coplin jar with 4x SSC/0.1% Tween for 30 min at 37°C; shake each 10 min or perform this step in a shaking water bath or a rocking platform in an incubator.
d) Let the cover slip slide down the slide.

e) Incubate the slide without the cover slip in the second pre-heated coplin jar with 4x SSC/Tween for 30 min at 37°C shaking or occasionally shaking it.
f) Dehydrate slides in coplin jars in 70%, 80% and 96% EtOH, 3 min each at RT and let them air-dry.
g) Incubate slides in coplin jar with fixative (methanol/acetic acid 3:1 v/v) for 30 min at RT and then let them air-dry.
h) Perform aging of the slide as described. After dehydration in 70, 80 and 96% ethanol, slides are ready to be re-used or stored at –20°C.
i) Prior to repeated denaturation, the RNase and pepsin treatment are not necessary anymore. However, to be sure that no previously hybridized DNA will remain on chromosomes, it is advisable to prolong denaturation time of the slide by 30 s - 1 min. Optionally, it is also possible to increase the temperature of denaturation slightly (this has to be tested).

Troubleshooting

Quality and spreading of chromosomes

This is a crucial factor influencing considerably the overall hybridization as well as the evaluation of the resulting pattern. Generally, the better the chromosome spreading, the better hybridization and its evaluation.

Aging

Slides that were not aged properly can result in: 1. weak or uneven DAPI staining and/or 2. a puffy appearance of chromosomes after hybridization and/or 3. a weak hybridization signal.

Nick translation

The most important factor influencing the time required for the reaction and the length of the DNA fragment is the proportion of dUTP conjugated to haptens.

Precipitation of pellet

In the case that the pellet has not precipitated, it is necessary to check whether the appropriate amount of EDTA has been used to stop the NT reaction (as indicated by the manufacturer of the commercial NT kits). Optionally, it is possible to prolong the freezing (–20°C) of the mixture of gDNA, EtOH, NaAc and salmon sperm for up to several days (when necessary and/or to improve the precipitation of the pellet).

Unspecific background signal

The bond between antigen and unspecifically attached hapten can result in an unspecific background signal. This can be reduced or avoided by increasing the concentration of BSA in the BSA/4x SSC solution or prolonged incubation with this solution at 37°C. As an alternative, 0.4 M NaCl can be used. Washing steps during detection are also crucial to get rid of unspecific background signals—the coplin jars have to be preheated to the desired temperature. If possible, use a shaking water bath or a rocking platform in an incubator set to the required temperature. Should this be impossible, at least occasionally gently shake the coplin jars with slides.

Laboratory set up

Equipment

Major

Agarose electrophoresis with documentation system; centrifuge; vortex; water bath; incubator/oven; mini-centrifuge; freezer (–20°C); epifluorescence microscope equipped with CCD camera; shaking water bath or rocking platform in an incubator (optional).

Minor

1.5 ml microtubes; coplin jars; microscopic slides with cover slips; nail polish; micropipette (1–1000 µl) with tips; fine forceps; timers.

Biochemicals and reagents

Ethanol 96%; formamide (Sigma-Aldrich); agarose in an electrophoresis buffer; dNTP-Mix (0.5 mM dATP, dCTP, dGTP+ 0.1 mM dTTP); 0.1 M mercaptoethanol (Merck); DNA polymerase I (Roche); DNase I (Roche; 100 mg diluted 1:250 in ddH_2O); biotin-16-dUTP (Roche); digoxigenin-11-dUTP (Roche); hapten succinimidyl esters (biotin, digoxigenin, dinitrophenol); aminoallyl-dUTP (Invitrogen); 1 M Tris-HCl pH 7.75; 2 M glycine pH 80.2 M sodium bicarbonate pH 8.3; BSA; PBS; SSC; Tween; 3 M natrium (sodium) Acetate; hybridization buffer; salmon sperm DNA; antifade (Vectashield); commercial kits: DNeasy Blood & Tissue Kit (Qiagen); nick translation mix for *in situ* probes (Roche).

FITC-streptavidin (Invitrogen) 1:20 to 1:100 dilution with PBS containing 10% FBS or dilution for specific application should be determined by the investigator to obtain the best conditions.

Cy™3-streptavidin (Invitrogen) 1:20 to 1:100 dilution with PBS containing 10% NGS.

Anti-digoxigenin-fluorescein, Fab fragments (Roche) 1:15 to 1:20 dilution with PBS containing 0.5% BSA.

Anti-digoxigenin-rhodamin, Fab fragments (Roche) 1:15 to 1:20 dilution with PBS containing 0.5% BSA.

Stock and working solutions

NT dNTP-Mix
0.5 mM dATP
0.5 mM dCTP
0.5 mM dGTP
0.1 mM dTTP
10x NT buffer
1 ml 1 M Tris-HCl, pH 7.5
100 ul 1 M $MgCl_2$
10 ul 10% BSA
890 ul ddH_2O
Store in 100 ul aliquots at –20°C.

Hybridization buffer (*sensu* Cremer et al. (2008))

5 ml formamide
2 ml 50% dextran sulfate
1 ml 20x SSC
400 µl 1M natrium-phosphate buffer
100 µl 10% SDS
200 µl 50x Denhardt's solution
1.3 ml ddH2O (total volume 10 ml), store at –20°C.

50x Denhardt's solution (or Sigma-Aldrich)

2 g BSA
2 g Ficoll (type 400)
2 g polyvinylpyrrolidone
200 ml ddH_2O
Filter to sterilize, divide into aliquots, and store at –20°C.

Pepsinization solution (0.005% pepsin in 0.0 1M HCl)

50 µl 10% pepsin
10 ml 0.1 M HCl
90 ml ddH_2O
Warm up to 37°C, store at –20°C.

RNase solution

10 µl RNase aliquot (25 mg/1.25 ml ddH_2O)
1 ml 2x SSC.

Acknowledgements

We thank Dr. Christian Lanctôt, PhD of the First Faculty of Medicine, Charles University in Prague for reviewing the draft of this protocol. This protocol has been tested and accomplished with contribution of projects no. P506/11/P596 and P523/08/0824 of the Czech Grant Agency. We also gratefully acknowledge the grant support provided by the Grant Agency of Charles University in Prague (GAUK) no. 158110.

References

Bi, K. and J.P. Bogart. 2006. Identification of intergenomic recombinations in unisexual salamanders of the genus *Ambystoma* by genomic *in situ* hybridization (GISH). Cytogenet. Genome Res. 112(3-4): 307–312.

Cremer, M., F. Grasser, Ch. Lanctôt, S. Müller, M. Neusser, R. Zinner, I. Solovei and T. Cremer. 2008. Multicolor 3D fluorescence *in situ* hybridization for imaging interphase chromosomes. pp. 205–239. *In*: Hancock, R. (ed.). The Nucleus: Volume I: Nuclei and Subnuclear Components, Methods in Molecular Biology™. Volume 463. Humana Press.

Fujiwara, A., C. Nishida-Umehara, T. Sakamoto, N. Okamoto, I. Nakayama and S. Abe. 2001. Improved fish lymphocyte culture for chromosome preparation. Genetica. 111: 77–89.

Graham, D.E. 1978. The isolation of high molecular weight DNA from whole organisms of large tissue masses. Anal. Biochem. 85: 609–613.

Henegariu, O., N.A. Heerema, L.L. Wright, P. Bray-Ward, D.C. Ward and G.H. Vance. 2001. Improvements in cytogenetic slide preparations: controlled chromosome spreading, chemical aging and gradual denaturing. Cytometry 43: 101–109.

Iourov, I. 2009. Microscopy and imaging systems. pp. 75–84. *In*: Liehr, T. (ed.). Fluorescence *In Situ* Hybridization (FISH)—Application Guide. Springer-Verlag, Berlin, Heidelberg.

Kallioniemi, A., O.P. Kallioniemi, D. Sudar, D. Rutovitz, J.W. Gray, F. Waldman and D. Pinkel. 1992. Comparative genomic hybridization for molecular cytogenetic analysis of solid tumors. Science 258(5083): 818–821.

Knytl, M., L. Kalous, R. Symonová, K. Rylková and P. Ráb. 2013. Chromosome studies of European cyprinid fishes: cross-species painting reveals natural allotetraploid origin of *Carassius* female with 206 chromosomes. Cytogenet. Genome Res. 139(4): 276–283.

Landegent, J.E., N. Jansen in de Wal, R.W. Dirks, F. Baao and M. van der Ploeg. 1987. Use of whole cosmid cloned genomic sequences for chromosomal localization by non-radioactive *in situ* hybridization. Hum. Genet. 77(4): 366–370.

Lim, K.I., R. Matyášek, A. Kovařík and A. Leitch. 2007. Parental origin and genome evolution in the allopolyploid *Iris versicolor*. Ann. Bot-London. 100: 219–224.

Matoba, H., A. Soejima and Y. Hoshi. 2007. Identification of parental genomes and genomic organization in *Aster microcephalus* var. Ovatus. J. Plant Res. 120: 585–593.

Ráb, P. and P. Roth. 1988. Cold-blooded vertebrates. pp. 115–124. *In*: Balicek, P., J. Forejt and J. Rubes (eds.). Methods of Chromosome Analysis. Cytogenet. Sect. Cs. Biol. Soc. Publishers, Brno.

Sealey, P.G., P.A. Whittaker and E.M. Southern. 1985. Removal of repeated sequences from hybridisation probes. Nucleic Acids Res. 13(6): 1905–22.

She, Ch.-W., Y.-Ch. Song and X.-H. Jiang. 2012. Comparative genomic *in situ* hybridization analysis on the chromosomes of five grass species with rice genomic DNA probe. Afr. J. Biotechnol. 11(39): 9508–9515.

Sommerauer, M. and I. Feuerbacher. 2009. Optical filters for FISH. pp. 85–98. *In*: Liehr, T. (ed.). Fluorescence *In Situ* Hybridization (FISH)—Application Guide. Springer-Verlag, Berlin, Heidelberg.

Symonová, R., M. Flajšhans, A. Sember, M. Havelka, D. Gela, T. Kořínková, M. Rodina, M. Rábová and P. Ráb. 2013a. Molecular cytogenetics in artificial hybrid and highly polyploid

sturgeons: an evolutionary story narrated by repetitive sequences. Cytogenet. Genome Res. 141(2-3): 153–162.
Symonová, R., Z. Majtánová, A. Sember, G.B.O. Staaks, J. Bohlen, J. Freyhof, M. Rábová and P. Ráb. 2013b. Genome differentiation in a species pair of coregonine fishes: an extremely rapid speciation driven by stress activated retrotransposons mediating extensive ribosomal DNA multiplications. BMC Evol. Biol. 13: 42.
Takahashi, C., J.A. Marshall, M.D. Bennett and I.J. Leitch. 1999. Genomic relationships between maize and its wild relatives. Genome 42: 1201–1207.
Traut, W. and H. Winking. 2001. Meiotic chromosomes and stages of sex chromosome evolution in fish: zebrafish, platyfish and guppy. Chrom. Res. 9(8): 659–672.
Traut, W., K. Sahara, T.D. Otto and F. Marec. 1999. Molecular differentiation of sex chromosomes probed by comparative genomic hybridization. Chromosoma. 108: 173–180.
Trifonov, V.A., N.N. Vorobieva and W. Rens. 2009. FISH with or without COT1 DNA. pp. 99–109. *In*: Liehr, T. (ed.). Fluorescence *In Situ* Hybridization (FISH)—Application Guide. Springer-Verlag, Berlin, Heidelberg.
Valente, G.T., C.H. Schneider, M.C. Gross, E. Feldberg and C. Martins. 2009. Comparative cytogenetics of cichlid fishes through genomic *in situ* hybridization (GISH) with emphasis on *Oreochromis niloticus*. Chrom. Res. 17: 791–799.
Walsh, P.S., D.A. Metzger and R. Higuchi. 1991. Chelex® 100 as a medium for simple extraction of DNA for PCR-based typing from forensic material. Biotechniques 10: 506–513.
Zhu, H.P. and J.F. Gui. 2007. Identification of genome organization in the unusual allotetraploid form of *Carassius auratus gibelio*. Aquaculture 265(1-4): 109–117.

17

Fish Chromosome Microdissection and its Utilization in the Synthesis of Fluorescence *in situ* Hybridization Probes

*A. Laudicina** and *M.C. Mühlmann†*

Molecular Cytogenetic Laboratory, National Commission of Atomic Energy, 1499 General Paz Av. San Martín (1650). Buenos Aires, Argentina.

*Corresponding author: alaudicina@lexelmedical.com

Principles

There are several critical steps required to obtain a chromosome-specific probe for fluorescence *in situ* hybridization (FISH). Initial DNA is obtained by mechanically microdissecting selected regions of target chromosome in metaphase spreads, followed by an unspecific Polymerase Chain Reaction (PCR) amplification and a subsequent regular stringency PCR to obtain quantitative amplification of the microdissected DNA. Finally, the DNA must be labelled in order to function as a FISH probe. Every step can be individually checked but only once the procedure is finished, in other words when the hybridization is developed on the slide, we can ascertain a successful procedure.

To microdissect such a small chromosome region, an inverted microscope with a 100x objective is needed along with an adequate micromanipulator enabling fine distance movement control of about 150 nm per step.

Microdissection needles also have a small size diameter at their tips, therefore a micropipette puller is recommended, to make capillaries with a diameter as small as 0.4 µm. Microdissected pieces are directly collected into the PCR tube where the first amplification reaction will occur.

The next step is a ramped, low annealing temperature PCR to amplify the dissected DNA. The principal characteristic of this PCR is the use of Degenerated Oligonucleotide Primers (DOP) which have 6 random bases. Using DOP makes the procedure useful for any species for which metaphase chromosome can be obtained. This PCR reaction has two stages; the first one has only 12 cycles and involves a very low annealing temperature which favors the annealing of primers at many random regions of the dissected DNA. If this step is successful, the original microdissected chromosome would be represented by thousands of fragments with DOP primers added in each end. Twenty five to thirty more cycles follows, as a quantitative amplification of these fragments. Here, regular parameters of annealing are used as the primers have already been incorporated in the amplified DNA.

Once the DNA has been successfully amplified a fluorescent label or hapten molecule should be incorporated into the DNA to be used as a detectable probe. This can be easily accomplished by a custom *in vitro* labelling reaction involving DNA polymerization where a tagged nucleotide replaces a proportion of the regular one. Sometimes a minor concentration adjustment might be necessary. Finally, if every step functioned correctly the DNA probe is ready for FISH.

Applications and Limitations

Although the first chromosomes microdissections were made in the late 70's (Hill and Watt 1977; 1978), the application of the technique to obtain FISH probes didn't begin till the 90's (Kao 1990; Meltzer et al. 1992). Several technological advances such as the possibility to amplify DNA *in vitro* improved the process efficiency and spread the use of the technique. Today, specific variations of PCR allow amplifying virtually any DNA present in the PCR mix, which permits scientists the possibility of working with unknown DNA sequences. So, the potential to synthesize probes from any organism introduced molecular cytogenetic analysis into many unexplored fields. There are various examples of probes for non-human organisms developed through this method, like insects (Staiber et al. 2000; Siirin et al. 2003), animals (Nieves et al. 2005; Ulsh et al. 2000) and plants (Shibata et al. 2002; Hizume et al. 2001). Home-produced probes expand the possibilities at the molecular biology level for unknown DNA segments and for many biological entities. This characteristic is especially important for those research groups whose biological models are not commercially relevant at the present time, hence making it almost impossible to get commercial specific probes. The raising number of published papers in which this technology is used shows its usefulness. In chromosome evolution studies, it has open the possibility of physically locate, cytogenetically, sequences of related species (Nieves et al. 2005) and follow structural changes of a particular chromosome through time line (Weise et al. 2005). These probes have also been used to determine the degree of chromosome relationship between species with different grades of "classical" divergence, functioning as a very accurate tool for systematics (Nieves et al. 2006). Biotechnology is another field where microdissected probes are increasingly used. Now it is possible to test chromosome constitution of cattle embryos, choosing which will be transferred to the recipient female (Di Berardino et al. 2004; Muhlmann et al. 2005). These are only a few examples of many found in the literature.

Theoretically, this technique allows the synthesis of probes for virtually any organism. However some details should be taken into account before starting. Probably the first point to handle is the quality of the metaphase spread to be microdissected. Any interference with the target could cause the failure of the whole process. A common occurence is the proximity of the chromosomes in the metaphase. The target chromosome or region should be as isolated as possible in order to avoid the contamination with non-specific material, which won't be discriminated in the first amplification step due to the low stringency conditions of the reaction. The presence of cytoplasm represents another problem related with the quality of the metaphases. Remaining cytoplasm acts as a "net" over the whole metaphase making it difficult to retrieve only the desired chromosome.

The exact identification of the target chromosome is critical as any error at this point, could cause the failure of the whole process. In fish karyotypes when chromosomes are not morphologically distinct by Giemsa staining, it is almost impossible to get the target DNA. As the dissected DNA will be in direct contact with the PCR polymerase, any interference with other reagents coming from differential cytogenetic techniques should be avoided.

Once the above details have been taken into account, it is necessary to evaluate the number of microdissections of the same target to collect. Although there are reports where FISH probes have been obtained with only one or two microdissected chromosomes (Guan et al. 1993; Viersbach et al. 1994; Christian et al. 1999), the general rule is that for the probability of success to be acceptable the quantity of chromosomes should be above the bare minimum. A good starting point for evaluating the number of microdissections necessary is to consider size and constitution of the target chromosome. Small chromosomes will have more chances of being entirely and homogeneously amplified in a PCR reaction, because it is more likely to get them all at once into a tube. On the contrary long chromosomes need, usually, more than one dissection to lift every piece. On the other hand, DNA probes from heterochromatic chromosomes rich in repetitive sequences, are often easier to obtain as any sequence amplified will hybridize along the entire heterochromatic block.

Regardless of microdissection and amplification performances, the real outcome is observed after the FISH technique. In about 10% of the procedures, clear and accurate microdissections followed by good amplifications (viewed on agarosa gels) gives no hybridization signals. On the other hand fluorescent signals might light the entire metaphase homogeneously with no differentiation of the target. Such results constitute another limitation of the technique.

It has to be considered that a single cell contaminating the PCR collection tube will introduce more DNA than the microdissected target; this could generate a probe for the whole karyotype (if the cell is of the same specie than the target) or a good amplification but no results (if a foreign cell is involved). To minimize this problem sterilization of all the material used in the procedure with UV light (330 nm) is recommended. The radiation will crosslink contaminated DNA inhibiting elongation which will otherwise take place.

Since this technology allows the amplification of any DNA sequence present in the PCR reaction, some considerations must be taken regarding chromatin constitution of the target chromosome: unique, and different amount of highly repetitive sequences characteristic of each species. When a chromosome is microdissected, amplified and used as a probe, repetitive sequences can cause a generalized hybridization where the target chromosomes are only slightly more fluorescent than the rest, making the analysis very difficult. Overcoming this issue is crucial in order to obtain a good quality DNA probe. To solve this problem a proportion of repetitive DNA (Cot DNA) in a higher concentration than the present in the probe should be present in the hybridization mix (Ulsh and Boreham 2003). In the right amount, a competitive reaction will block general repetitive sequences resulting in specific signals from the target itself.

Protocol

1. Coverslip preparation of metaphase spreads

The microdissector is mounted on an inverted microscope (Fig. 1A); the objectives will be positioned under the slide and the microdissector needle on top where the metaphases are found. Minimum objective requirements are: 10x to locate the metaphases and 100X to perform the microdissection.

The focus distance of objective to the target is shorter than a regular slide width; therefore the metaphases should be spread on a coverslip (Guan et al. 1993; Muhlmann-Díaz et al. 1995).

a) From a freshly prepared standard cytogenetic cellular suspension, drop 24 x 50 mm coverslips previously cleaned.
b) Stain with 10% Giemsa for 10 minutes.

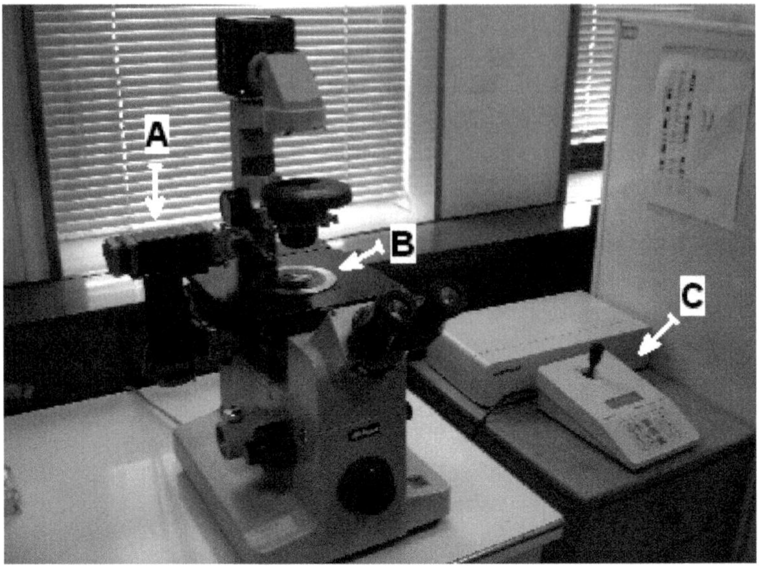

Figure 1. Inverted microscope equipped with microdissector. Arrows indicate: (A) microdissector arms, (B) adapted rotary stage, (C) microdissector joystick.

c) Rinse with distilled water and air dry.

2. Microdissection needles

The required needle tip diameter varies with the size of the target. Large and whole chromosomes will be easily microdissected with wider needle tips while smaller ones and parts (centromeres, telomeres, etc.) need thinner tips. The capillaries used to obtain the needles are thin wall borosilicate glass with 1 mm exterior diameter/0,75 mm interior diameter. Different puller brands are available to pull the micro-pipette to the right size.

a) In accordance with the microdissection target set up the heater of the pipette puller. A higher heat level will produce a thinner needle tip diameter.
b) Store the needles until use in an appropriate environment to avoid dust, contamination or breakage.

3. Microdissection

This protocol is based on a microdissector that has a minimum movement limit of 150 nm, and SLOW and FAST modes that according to joystick deflection, increases the speed progressively from 0 to 700 mm/seg and to 5500 mm/seg respectively.

The inverted microscope, where the microdissector is mounted, should have two rotating stages. One on top of the other and with a grease substance (like solid Vaseline) between them (Fig. 1B). Working with both stages allows the tight orientation and location of the target chromosome. When a whole

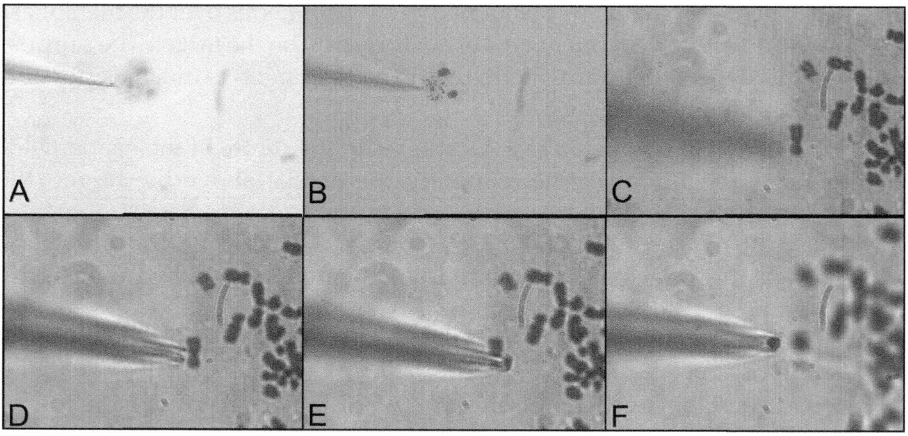

Figure 2. Different stages of microdissection process. (A-D) Second stage approximation, (E) Microdissection, (F) Microdissected material into the needle tip.

chromosome will be microdissected, a parallel orientation is recommended; when working with parts of chromosomes it's better to rotate the chromosome perpendicular to the needle (Fig. 2D).

In order to avoid vibration and needle tip breakage (Fig. 1C) it is very important to place the control joystick of the microdissector on a different table than the microscope with the micromanipulator.

Gloves must be used during the whole process to minimize contamination.

Care must be taken to assure the cleanness of the area where the microdissection takes place. It is recommended that the microscope's general parts, furniture, racks, etc. be previously cleaned with ethanol 70%.

a) Place coverslip on the stage. Check that the cellular spread is upwards.
b) Find a good metaphase where the target chromosome can be clearly identified.
c) If necessary, rotate the stage to place the target in a suitable orientation.
d) Focus the chromosome of interest with the 100x objective.
e) First approximation:
    - Engage the needle in the holder and adjust to the microdissector arm.
    - In FAST mode, approach the needle in the X and Y-axis to the lens of the 100x objective.

- Change to SLOW mode and complete the approximation in the Z-axis as slowly as possible. Stop Z movement when the (visible) tip of the needle is about 3 mm from the coverslip surface.
- Change the objective to 10x and focus on the metaphase.

f) Second (final) approximation:
The needle tip should be positioned a few millimetres over the metaphase in the objective field but not necessarily exactly on the target. The purpose of this stage is to approximate the tip to the target as much as possible but without making contact.

- Ensure that the metaphase location is in the centre of the vision field.
- Raise the 10x objective focussing the needle above the target (the metaphases will loose focus).
- Locate the needle on the centre of the vision field where the metaphase is. At this time the needlepoint is a few millimetres above and exactly over the metaphase.
- Descend the objective towards the metaphase until the needlepoint is out of focus.
- Descend the needlepoint until it is within the focus plan again.
- Repeat the two previous steps enough times to approximate the needlepoint to a few micrometers above the metaphase (at this point a shadow of the needle appears over the focused metaphase) (Fig. 2A).
- Focus on the metaphase, now the needle will appear as a blurred silhouette (Fig. 2B).
- Change the objective to 100X and re-focus on the metaphase. This is the last opportunity to adjust any parameter from the microscope (light, focus, etc.).

g) From this step, special care should be taken to avoid vibrations (do not touch the microscope or anything else but the joystick controlling the micromanipulator) as any little vibration can break the needle tip.

- The last approximation step is a very slow movement down. The needle, at this time will be very close but out of focus. At this point, only the needle shadow will be seen, but while the needle goes down the focus will improve (Fig. 2C).
- When the metaphase and needlepoint get in the same visual plane, place the tip within a tiny distance ahead of the target and continue moving down (Fig. 2D).
- The moment the tip touches the coverslip is marked by the fact that the needle begins to move forward instead of downward (This is because the needle tip bends as it touches the coverslip surface).
- Now just move the needle across the target, if everything works properly the DNA should enter the needle (Fig. 2E).

h) To remove the needle with the dissected material it is recommended to proceed with a forward and upwards movement (X and Z-axis) at the same time.

- As soon as the needle is no longer in contact with the slide, check if the dissected material is inside it by focussing gently (Fig. 2F).
- Select FAST mode, raise the needle and disengage from the holder.
- Break the needle tip into the very bottom of a sterile PCR tube. This tube will contain the PCR reaction mix recommended by Telenius et al. (1992). Repeat steps b) to h) several times (10 to 20 pieces if possible).

4. Post-microdissection treatment

a) Heat the mix containing the dissected chromosomes to 90°C for 10 min.
b) Spin down the solution to avoid evaporation.
c) Add 2 µl thermosequenase (4U μl^{-1}).

5. First PCR reaction (RAMP-PCR)

This is a two in one reaction; the tube must not be removed from the PCR machine until the 35th cycle has been completed.

A. Set the PCR machine with the following parameters:
   a) 95°C 3 min
   b) 94°C 1.5 min
   c) 30°C 3 min
   d) Ramp to 72°C at approximately 0.2°C seg^{-1}
   e) 72°C 1.5 min
   f) Repeat step b) through 5:10 times
   g) 94°C 1.5 min
   h) 56°C 1.5 min
   i) 72°C 1.5 min
   j) Repeat steps 7 through 9:35 times.

B. Run a PCR of the dissected material in the collection drop with this program.
C. Check the result on a 1% agarose gel with a 100 bp standard (usually with the above PCR settings a 200–600 pb smear is obtained).
D. Store at –20°C.

6. Second PCR reaction (stock)

This is a standard reaction. The chromosome fragments amplified in the first PCR have already incorporated the Universal primers and no longer need low astringency settings. The purpose of this reaction is to obtain stock DNA to use as template in labeling reactions, and to avoid running out of originally dissected-amplified material.

A. In a sterile PCR tube mix the next reagents: 33.5 µl UV sterilized water, 5 µl of 10x PCR reaction buffer, 4 µl of $MgCl_2$ (25mM) of dNTP stock (2.5 mM each), 3 µl DOP primer (100 µM) and 0,5 µl of Taq polymerase.
B. Add 2 µl of the 1° PCR product.
C. Set the PCR machine with the following parameters:
   a) 94ºC 2.5 min
   b) 94ºC 1.5 min
   c) 56ºC 1.5 min
   d) 72ºC 1.5 min
   e) Repeat steps b) to d), 30 times
   f) 72ºC 2 min
D. Run this PCR program on the mix prepared above.
E. Check the result on a 1% agarosa gel with a 100 bp standard (usually with the above PCR settings a 300–500 pb smear is obtained).
F. Store at –20ºC.

7. Third PCR reaction (labeling)

Although several *in vitro* reactions are used to label DNA, our laboratory utilizes PCR with the purpose of not only labeling but also amplifying the target.

The protocol used is a modification of Bray-Ward (2002).

The procedure is the same as for the 2° PCR. The single difference is that only 1.25 mM of dTTP is used and 20 µM of a dUTP tagged with a reporter molecule is added to the PCR mix.

After the reaction, check the result on a 1% agarose gel with a 100 bp standard (usually with the PCR settings used, a 300–500 pb smear is obtained).

8. DNA precipitation

Sometimes, a precipitation step is necessary in order to concentrate the DNA. The protocol used is a modification of Sambrook et al. (1989).
To the final volume of PCR add 2/3 vol. of 5 M Ammonium acetate.

a) Add 3 volumes of ice cold 100% Ethanol
b) Store for a minimum of 30 minutes at –20°C or –70°C
c) Centrifuge 15 minutes at 10000 g
d) Carefully decant supernatant
e) Dry the pellet

f) Dissolve the pellet in TE buffer (Ph:7) or Ultra pure water, to a final concentration of 0.1 µg/ml
g) Store at –20°C

Troubleshooting

Some problems may be encountered through the application of this technique:

Microdissection

- The dissected material doesn't enter into the needle tip due to static electricity: a steam source should be positioned near the microdissector in order to moisturize the ambient.
- Surrounding chromosomes are lifted up together with the target (this effect is a common consequence of cytoplasm debris over the metaphase): improve metaphase spreads, no cytoplasm should remain and the chromosomes should be correctly separated.

First PCR reaction (RAMP-PCR)

- PCR products are obtained but FISH reactions are negatives: the possible cause of this is PCR contamination: increase asepsis rules. Include negative control tubes with PCR mix but no DNA.
- No PCR product: check PCR reagents.
- No PCR product: more microdissected pieces are needed.
- Low amplification level: although this is a not desired outcome, sometimes it is a common result. It is recommended that the whole process be continued and check the results on a FISH reaction.

Equipment

Major

Microdissector (minimum movement limit of 150 nm); inverted microscope; pipette puller, micropipettes; thermocycler (with ramp capabilities); micro centrifuge; electrophoresis: cube and power source; freezer –80°C.

Minor

Coverslips (24 x 50 mm); PCR tubes; thin wall borosilicate glass capillaries; tips; gloves.

Biochemicals and reagents

Giemsa; agarose; Taq polymerase; sequenase; dATP; dCTP; dGTP, dTTP; tagged dUTP; polymerase 10x buffer; sequenase 10x buffer; ethanol; ammonium acetate; DNA ladder (100 bp); ethidium bromide.

Stock and working solutions

Telenius (1992) PCR reaction mixture: 10 µl UV sterilized water, 2 µl of 10x Thermosequenase reaction buffer, 4 µl of dNTP stock (2.5 mM each) and 2 µl DOP primer solution, 5′ CCG ACT CGA GNN NNN NAT GTG G 3′ (25 µM).

References

Bray-Ward, P. 2002. FISH probes and labelling techniques. pp. 5–28. *In*: Beatty, B., S. Mai and J. Squire (eds.). FISH. A Practical Approach. Oxford University Press, Oxford, New York.

Christian, A.T., H.E. Garcia and J.D. Tucker. 1999. PCR *in situ* followed by microdissection allows whole chromosome painting probes to be made from single microdissected chromosomes. Mamm. Genome 10(6): 628–31.

Di Berardino, D., M. Vozdova, S. Kubickova, H. Cernohorska, G. Coppola, G. Enne and J. Rubes. 2004. Sexing river buffalo (*Bubalus bubalis* L.), sheep (*Ovis aries* L.), goat (*Capra hircus* L.), and cattle spermatozoa by double color FISH using bovine (*Bos taurus* L.) X- and Y-painting probes. Mol. Reprod. Dev. 67(1): 108–115.

Guan, X.Y., J.M. Trent and P.S. Meltzer. 1993. Generation of band-specific painting probes from a single microdissected chromosome. Hum. Mol. Genet. 2(8): 1117–1121.

Hill, R.J. and F. Watt. 1977. Chromosomal proteins of *Drosophila melanogaster* and an approach for their localisation on polytene chromosomes. Chromosoma. 63: 57–58.

Hill, R.J. and F. Watt. 1978. 'Native' salivary chromosomes of *Drosophila melanogaster*. Cold Spring Harbor Symp. quant. Biol. 42: 850–865.

Hizume, M., F. Shibata, Y. Maruyama and T. Kondo. 2001. Cloning of DNA sequences localized on proximal fluorescent chromosome bands by microdissection in *Pinus densiflora* Sieb. & Zucc. Chromosoma. 110(5): 345–351.

Kao, F.T. 1990. Microdissection and microcloning of human chromosome 21. Proc. Clin. Biol. Res. 360: 89–104.

Meltzer, P.S., X.Y. Guan, A. Burgess and J.M. Trent. 1992. Rapid generation of region specific probes by chromosome microdissection and their application. Nat. Genet. 1(1): 24–28.

Muhlmann, M.C., A.O. Laudicina, C. Perandones, M.V. Bertolino, A. Marazzi, C.J. Quintans, M. Donaldson, W. Bozzo and S. Pasqualini. 2005. Uses and limitations of two molecular cytogenetic techniques for the study of arrested embryos obtained through assisted reproduction technology. Genet. Mol. Res. 4(2): 143–151.

Nieves, M., M. Muhlmann and M.D. Mudry. 2005. Heterochromatin and chromosome evolution: a FISH probe of *Cebus apella paraguayanus* (Primate: Platyrrhini) developed by chromosome microdissection. Genet. Mol. Res. 30; 4(4): 675–83.

Nieves, M., A. Ortiz, G. Mendez, M. Muhlmann and M.D. Mudry. 2006. Molecular cytogenetics in Platyrrhini hybrids diagnosis: Conservation programs application. Int. J. Primatol. Vol. 27 Supplement 1. Feb. 2006. Program for the Twenty-First Congress of the International Primatological Society. Abstract #333.

Sambrook, J., E.F. Fritsch and T. Maniatis. 1989. Concentrating nucleic acids. pp. E10–E16. *In*: Sambrook, J., E.F. Fritsch and T. Maniatis (eds.). Molecular Cloning. A Laboratory Manual. Second Edition. Cold Spring Harbor Laboratory Press, USA.

Shibata, F. and M. Hizume. 2002. Evolution of 5S rDNA units and their chromosomal localization in *Allium cepa* and *Allium schoenoprasum* revealed by microdissection and FISH. Theor. Appl. Genet. 105(2-3): 167–172. Epub 2002 Jun. 21.

Siirin, M.T., N.B. Rubtsov, T.V. Karamysheva, A.V. Katokhin, D.A. Karagodin and I.I. Kiknadze. 2003. Molecular-cytogenetic characteristics of B-chromosomes in chironomid (Diptera, Chironomidae). Tsitologiia. 45(6): 582–589.

Staiber, W. and C. Schiffkowski. 2000. Structural evolution of the germ line-limited chromosomes in *Acricotopus*. Chromosoma. 109(5): 343–349.

Ulsh, B.A. and D.R. Boreham. 2003. *In vitro* biomarkers of radiation exposure in the environment. pp. 282–288. *In*: Carmel, M. and A. Brian (eds.). *In Vitro* Methods in Aquatic Ecotoxicology. Springer-Verlag, New York.

Ulsh, B.A., M.C. Muhlmann-Diaz, F.W. Whicker, T.G. Hinton, J.D. Congdon and J.S. Bedford. 2000. Chromosome translocations in turtles: a biomarker in a sentinel animal for ecological dosimetry. Radiat. Res. 153(6): 752–759.

Viersbach, R., G. Schwanitz and M.M. Nothen. 1994. Delineation of marker chromosomes by reverse chromosome painting using only a small number of DOP-PCR amplified microdissected chromosomes. Hum. Genet. 93(6): 663–667.

Weise, A., H. Starke, K. Mrasek, U. Claussen and T. Liehr. 2005. New insights into the evolution of chromosome 1. Cytogenet. Genome Res. 108(1-3): 217–222.

18

Microdissection and Chromosome Painting

M.R. Vicari,[1] *J.C. Pansonato-Alves*[2,a] and *F. Foresti*[2,b,*]

[1]Departamento de Biologia Estrutural, Molecular e Genética, Laboratório de Citogenética e Evolução, Universidade Estadual de Ponta Grossa, Av. Carlos Cavalcanti, 4748, 84030-900, Ponta Grossa, Paraná State, Brazil.
e-mail: vicarimr@pq.cnpq.br

[2]Departamento de Morfologia, Laboratório de Biologia e Genética de Peixes, Universidade Estadual Paulista, Distrito de Rubião Júnior, s/n, 18618-970, Botucatu, São Paulo State, Brazil.
[a]e-mail: pansonato@ibb.unesp.br
[b]e-mail: fforesti@ibb.unesp.br
*Corresponding author

Principles

1. Chromosome microdissection and amplification

Chromosome microdissection was originally developed in studies involving *Drosophila* polytene chromosomes as a way to construct DNA markers from specific chromosomal regions (Scalenghe et al. 1981). This methodology was first used to obtain clones of individual bands of polytene chromosomes and became known as microdissection and micro-cloning. This technique was subsequently used for obtaining markers for rat chromosomes and then extended to human chromosomes (Röhme et al. 1984; Bates et al. 1986). Nowadays, microdissection has been extensively used in comparative chromosome painting procedures, as well as for the obtainment of products for DNA sequencing (Vicari et al. 2010; Chen et al. 2012).

The principle of chromosome microdissection technique is based on the identification of either a chromosome, chromosome arm or chromosome band for the subsequent removal of such target with the aid of a micromanipulator attached to an inverted microscope. Once the target is defined, whole chromosomes, or chromosomal regions, are directly isolated under an inverted microscope linked to a micromanipulator. Following dissection, chromosomal fragments pass through a step of amplification, using adapted methods of whole genome amplification (WGA) in order to increase the limited amount of DNA (Telenius et al. 1992; Zhang et al. 1992) and reduce the starting copies material (Fig. 1).

Until now, different WGA techniques have been developed and the most widely used in microdissection assays is the degenerate oligonucleotide-primed-PCR (DOP-PCR), method that was originally developed by Telenius et al. (1992) and has been modified for amplification of microdissected chromosomes (Guan et al. 1992). This method uses a single primer that contains an *Xho*I restriction site at its 5′-end, and 3′-end, and a random hexamer sequence between them. The DOP-PCR comprises two different stages. In the first stage (low stringency) low annealing temperature is used to facilitate primer pairing in several sites in the target genome. The 3′-end of the primer binds to sites in the genome complementary to the 6-bp well-defined sequence at the 3′-end of the primer. The adjacent random hexamer sequence (displaying all possible combinations of the nucleotides A, G, C, and T) can then anneal and tag these sequences with the DOP primer. In the second stage, high stringency is employed and the product generated in the first cycle is amplified with a higher binding specificity of the primer. DOP-PCR generates smears of DNA fragments (200 to 600 bp), evidencing a drag on 1% agarose gel electrophoresis (Fig. 2). After that, a secondary DOP-PCR is carried out with the high stringency cycle to increase the amount of DNA.

The efficiency of a WGA method is measured by the coverage of the amplified material from the target. Although the DOP-PCR primer presents a degenerated structure and the first amplification step is based on low stringency, it may sometimes be difficult to obtain a complete coverage of

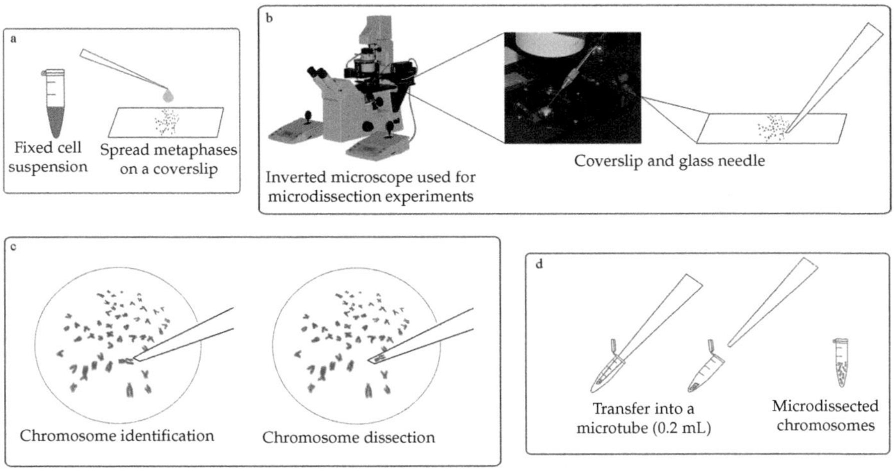

Figure 1. Representative schematic of the method for isolation of chromosomes by microdissection.

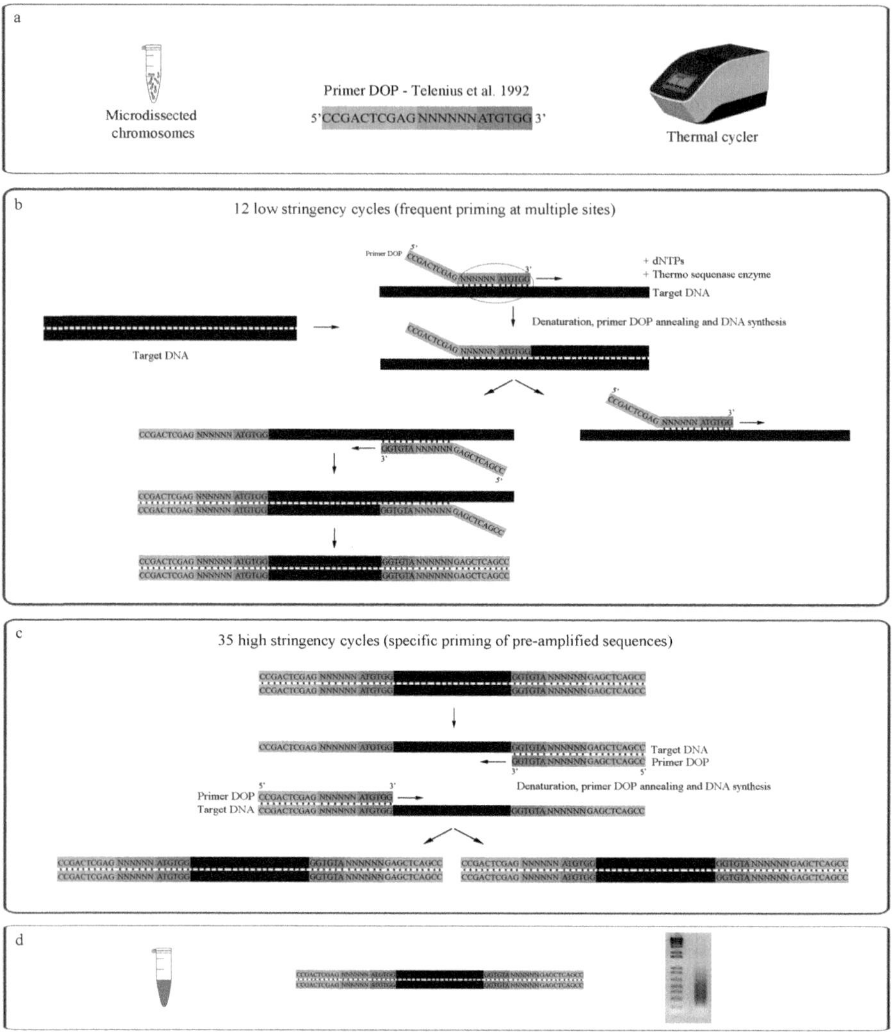

Figure 2. Amplification of microdissected chromosomes by DOP-PCR.

the microdissected chromosome. This fact may have many reasons, among which is a low number of target sequences for the DOP primer in the material to be amplified, which may be common in chromosomes rich in single-copy DNA. Furthermore, this technique normally results in incomplete genome coverage and amplification bias, in which a sequence is over represented in the amplified DNA due to preferential binding of the primers to specific regions. Furthermore, secondary DNA structures may adversely affect the success of a DOP-PCR primer annealing and therefore the amplification causing enzyme slippage or dissociation of the enzyme from the template giving origin to nonspecific amplification products, incomplete coverage of loci, and short DNA fragments (less than 100 bp) that cannot be used in DNA sequence applications.

As an alternative to the DOP-PCR, other methods, such as Genomeplex (Gribble et al. 2004) and GenomiPhi (Höckner et al. 2009), may be used for the amplification of microdissected chromosomes. These methods are based on the multiple displacement amplification (MDA) and involve the binding of random hexamers to denatured DNA followed by strand displacement synthesis at a constant temperature using the Phi29 DNA polymerase enzyme. Additional priming events can occur on each displaced strand leading to a network of branched DNA structures. Some of these methods, such as Repli-G (QIAGEN), illustra GenomiPhi DNA Amplification Kit (GE Healthcare Life Sciences), and Genome Plex Single Cell Whole Genome Amplification Kit (Sigma-Aldrich) are also available in commercial kits.

All those WGA procedures have been greatly used in various fields of genetics and research, such as genotyping, DNA sequencing, microarrays, microsatellite analysis, single nucleotide polymorphism (SNP) analysis, comparative genomic hybridization (CGH), and whole chromosome painting (WCP) (Peng et al. 2003; Lovmar et al. 2003; Bannai et al. 2004; Rook et al. 2004; Machado et al. 2011; Traut et al. 2013). Nevertheless, cytogeneticists use WGA methods, traditionally the DOP-PCR method, for synthesize probes obtained from microdissected chromosomes with emphasis on WCP (Vicari et al. 2010; Weise et al. 2010).

1.2 Painting probe production and Fluorescence *in situ* hybridization

The principle of chromosome painting is based on the synthesis of probes from whole, or chromosomal regions, obtained by microdissection or by Fluorescent Activated Chromosome Sorting (FACS), a subsequent amplification of the target DNA by a WGA method, followed by the labeling of that target DNA during secondary WGA reactions through the incorporation of modified nucleotides (e.g., Digoxigenin-11-dUTP, Tetramethyl-Rhodamine-5-dUTP). In this sense, the probes must be able to locate the chromosomal region from which it was taken by forming hybrid regions (probe-chromosomal DNA) and be detected using specific fluorescent-conjugated antibodies (e.g., Anti-Digoxigenin-Rhodamine). Thereafter, allowing the visualization of the relative position of the probes on the chromosomes. Thus, a good probe for chromosome painting must have full coverage of the target region or chromosome from which it was removed, be properly labeled with a fluorescent molecule or possess an antigenic site, and be synthesized in fragments from 100–500 bp for a better access to the chromosome structure during FISH.

Applications and Limitations

Chromosome microdissection allows the visualization, location, and isolation of DNA selected regions within a genome, which can be used in a wide range of applications. Once the isolation of the chromosome or chromosome segment is achieved, (i) it is possible to produce specific probes for the accomplishment of chromosome painting in genome differentiation and karyotype evolution studies; (ii) obtain repetitive DNA probes for physical location on the chromosome; (iii) apply the next generation sequencing (NGS) technology; and construct chromosome sequence and/or linkage groups to be assigned to specific chromosomes.

The development of FISH allowed the direct visualization and localization of selected DNA regions within a genome. Thus, it is not surprising that FISH was rapidly introduced for a variety of applications. FISH has been employed in both biology and medicine for karyotype analysis, study of chromatin organization, and other applications (Chen et al. 1993; Lai et al. 2000; Volpi et al. 2000; Chambeyron and Bickmore 2004). Comparative cytogenetic studies have traditionally relied on the examination of chromosome banding patterns. However, when considering species with highly rearranged genomes or with highly diverged species, cytogenetic comparisons of banding patterns are inadequate (Chowdhary and Raudsepp 2001). Hence, comparative chromosome painting associated with classical chromosomal markers has become the method of choice for performing cytogenetic genome comparisons. The advantage of such a method is that it allows for the identification of homologous genomic regions between species through cross-species *in situ* hybridization of whole chromosomes (Ráb et al. 2008).

In order to locate target regions, microdissection followed by chromosome painting procedures is widely used and have been useful to identify the origin of chromosome abnormalities in rearranged human karyotypes or to detect and diagnose double minute chromosomes in cancer tumors, among others (Kim et al. 2001; Arens et al. 2004; Vermeesch et al. 2005; Mahjoubi et al. 2006). In other groups of organisms, chromosome painting has been used to identify the origin of B chromosomes (Teruel et al. 2009; Voltolin et al. 2010; Vicari et al. 2011); the origin and differentiation of heteromorphic sex chromosomes in groups of adjacent organisms (Machado et al. 2011; Schemberger et al. 2011; Hirai et al. 2012); and in studies of genome rearrangements by Zoo-FISH (Rens et al. 2007; Ferguson-Smith and Trifonov 2007).

Chromosome microdissection and chromosome painting have been used in an attempt to elucidate the origin and composition of B chromosomes in some species, such as *Prochilodus lineatus* (Voltolin et al. 2010), *A. scabripinnis* (Vicari et al. 2011) and *Lithochromisrubripinnis* (Yoshida et al. 2011). However, over time B chromosomes differ quantitatively by evolution of sequences and qualitatively through chromosomal rearrangements. Due to the absence of selection pressure, mutations accumulate and recombination is prevented. These events simply indicate that there is a loss of functional loci on B chromosomes (Green 1990).

Due to all this structural and sequence differentiation in supernumeraries, painting of highly differentiated B chromosomes tend to have little or no recognition of sites in the A complement, and when visible signs are detected, they are usually regions containing repeated sequences.

The origin of sex chromosome systems in non-mammalian organisms is another matter widely studied using microdissection and comparative chromosome painting. At the origin of sex chromosomes, two main factor said in the differentiation between the proto sex chromosome pairs through the loss of recombination: (i) the amplification of genetic modifiers and (ii) chromosomal rearrangements. Rearrangements, such as translocations and inversions cause the reduction or suppression of crossing over in regions close to the breaking point. In those regions where crossing over was reduced or absent, there is a tendency towards accumulation of satellite DNA sequences in tandem, as well as of transposable elements. Regions of satellite DNA accumulation (heterochromatin) are known for the absence of crossing-over. Due to insertions of these elements and satellite DNA, the neighboring regions adjacent to the region without recombination would also be restricted to crossing-over (Charlesworth et al. 2005). In this connection, the generation of probes for painting obtained from differentiated sex chromosomes provides evidence for the occurrence of cross-species hybridization regions of the proto sex chromosome pair in those species that do not present sex chromosome heteromorphism.

Particularly in the group of fishes, the evolutionary lability and alternatives of sex-determining systems is remarkable (Mank et al. 2006). Considering only the Neotropical genus *Eigenmannia*, at least three different sex chromosomes systems were described (XX/XY, ZZ/ZW, $X_1X_1X_2X_2/X_1X_2Y$). In this sense, chromosome painting studies revealed that sex chromosomes are not homologous among each other in different species and may even be fused with another autosomes constituting the autosomal complement in sister species (Henning et al. 2011). This approach has been used in several groups of fishes and helped to shed light on genetic origin and evolution of distinctive sex chromosomes (Diniz et al. 2008; Cioffi et al. 2011a; 2011b; Machado et al. 2011; Pazian et al. 2013).

Evolutionary chromosome rearrangements can be considered 'rare genomic changes' (Rokas and Holland 2000) with very low levels of convergence. Hence, the observation of identifiable homologous chromosomes and chromosome segments represents a macroevent in an evolutionary lineage. The synthesis of probes for chromosome painting provides new opportunities for the analysis of the differentiation of group chromosomes. Thus, chromosome painting allows for cross-species hybridization, which has shown that the synteny for large genomic segments has been conserved in several mammals (Beklemisheva et al. 2011; Picone et al. 2011) and other groups (Nanda et al. 2011; Biltueva et al. 2011). However, studies of this nature generally require obtaining whole chromosome-specific probes by Fluorescent Activated Chromosome Sorting (FACS), which could explain the lack of reports in fishes, since some restrictions such as: absence of compartmentalization of their genomes and the difficulty to obtain fibroblast cultures limits its application for establishing complete comparative chromosome maps in this group.

In a pioneer study on *Gymnotus* (Gymnotiformes, Gymnotidae) fish species, Nagamachi et al. (2010) used whole chromosomes probes prepared by FACS and used in cross-species hybridizations, showing the occurrence of multiple rearrangements in species whose only macro detectable rearrangement was a single chromosome fusion. Although technical limitations restrict studies like this, one should note that FISH using whole chromosome probes is a powerful tool for investigating fish karyotype evolution.

In the past years, there has been great progress in the genetic mapping of the genomes and organisms considered as models for study. Nevertheless, in contradiction to this major advance, the genetic markers or repetitive DNA sequences in the genome of most groups of organisms are scarce or no existent. In these groups, chromosome microdissection and fragment cloning may be considered a quick and efficient choice for the generation of probes, especially those of repetitive DNA elements (Vicari et al. 2010). Clones of these libraries specific to a region or chromosome band can be used in conjunction with the existing markers for the construction of genetic linkage maps and for an easier physical location of the repetitive elements, providing an important tool for understanding the chromosome structure and the organization of regions specific to the genome (Zhou and Hu 2007). In addition, physical maps of specific chromosome regions combined with chromosome walking can be built using either the chromosome library or the libraries of specific chromosome regions for the selection of cosmid and plasmid libraries, yeast artificial chromosomes (YACs) and bacterial artificial chromosomes (BACs).

More recently, the development of NGS technologies has provided innumerous insights into the organization of genomes (Metzker 2010). Cytogeneticists see the NGS as a powerful tool in data generation for the reconstruction of chromosome scaffolds and chromosome maps starting from specific regions (Chen et al. 2012). Countless groups of cytogeneticists apply the NGS technologies in an attempt to understand the organization and composition of sex and B chromosomes, trying to relate this knowledge to the understanding of the genetic differentiation of sex in the most diverse groups of non-mammals and to the possible roles regarding the presence of B chromosomes in the genomes of many species. The NGS has proven to be very effective at reconstructing single-copy euchromatic genes, whereas repetitive regions remain a major challenge. Thus, the assembly process of the DNA repeat segment remains questionable for both repeats and segmental duplications, as high sequence homogeneity among copies of a given repeat or duplication limit the potential to reconstruct sequence order (Mardis 2008; Alkan et al. 2010). This methodology represents a difficulty for reconstructing large scaffolds from contigs (Green 2001) and complicates gene assembly and annotation (Alkan et al. 2010). However, new experimental approaches for the assembly of repeats and duplications have been developed and promise to help resolve this methodological limitation (Metzker 2010).

Although microdissection and chromosome painting showed to be useful in evolutionary and applicable studies in fish, microdissetion shows some difficulties and limitations, among which are (i) significant amounts of depurination in chromosome preparations fixed in methanol/acetic acid (3:1), (ii) problems with contamination of the sample, and (iii) difficulty in accurately recognizing the chromosome or target region to be microdissected.

To prevent the chromosome depurination caused by acid treatment during the fixation of the samples, it is recommended that, after being duly fixed, the chromosome preparations be stored using only methanol at –80°C. As to the problems with contamination, it is recommended to start the procedures with the sterilization of the utensils and reagents that are used for chromosome preparations until the final stage of microdissection itself. During the microdissection stage, coverslips, microtubes, tips, staining reagents and capillary for the micromanipulator duly autoclaved must be used. Researchers must also wear safe-handling devices, such as gloves and mask and use the laminar flow biological safety cabinet for the manipulation of the microdissected material. In most species of non-mammalian vertebrates, the small chromosome compartmentalization and the difficulty in identifying the chromosome to be obtained on metaphase chromosomes of uniform size and morphology has been the biggest problem for the recognition of the target to be microdissected. Due to the small longitudinal compartmentalization of chromosomes in these organisms, banding methods can also result in poor resolution. Vicari et al. (2010) adapted a protocol for chromosome painting in Neotropical fishes using C-banding when the target chromosome has a heterochromatic band marker. However, C-banding should be as mild as possible to avoid degradation and extensive loss of chromatin.

Protocol

1. Chromosome microdissection

Chromosome preparations

To obtain the target chromosome or chromosome region, conduct your laboratory standard procedures for chromosome preparations with all utensils and reagents properly autoclaved. After the last stage of fixation and subsequent centrifugation of the material, discard the supernatant and add only methanol to the material. Store preferably in an ultra-freezer.

Cover slip preparation

Before initiating the procedure, clean 24 x 60 mm coverslips with alcohol and arrange a package of coverslips separated by brown paper. Place the package in the autoclave at maximum 121°C for 20 min. Having the coverslip ready, take a portion of the chromosome preparation stored in methanol and add three parts of acetic acid. Homogenize the chromosome preparation and drop the 24 x 60 mm coverslip with a distilled water layer heated at 60°C. Perform conventional staining with Giemsa 5% properly autoclaved.

Preparation of needles

For the preparation of the needles, use 1 mm outer diameter x 0.4 inner diameter capillary tubes duly autoclaved. Take the capillary tubes to the micropipette puller and adjust the temperature to prepare tips of approximately 0.7 μm for microdissection of whole chromosomes. To remove small chromosomes or chromosome bands, adjust the temperature of the micropipette puller to obtain thinner tips.

Microdissection in mechanical manipulator

Place the coverslip containing the chromosome preparation on the inverted microscope equipped with mechanical microdissector. Use the recently prepared needles to microdissect about 20 chromosomes or the chromosomal region of interest. Transfer the chromosomes into a sterile microtube for further WGA completion (Fig. 1).

2. Whole genome amplification from Microdissected Products (Figs. 2 and 3)

DOP-PCR first reaction

When using a laminar flow cabinet, handling a tube with the needle tips containing the target microdissected chromosomes and working on another without a sample (negative control), prepare the following amplification mixture:

- Thermosequenase buffer (1x concentrate);
- 2 μM primer DOP (5′-CCGACTCGAGNNNNNNATGTGG-3′);
- 200 mM dNTPs;
- add sterile water to a final volume of 20 μl;
- heat in thermo block or water bath for 15 min at 100°C;
- Mix and centrifuge mildly;
- add 2 μL of thermosequenase enzyme (5 U/μL);
- place the mixture in the thermocycler.

Reaction—First DOP-PCR

94°C for 5 min
12 low stringency cycles:

94°C for 1 min and 30 s
35°C for 10 s
Ramp: from 35°C to 72°C increasing 0.2°C/sec (approximately 2 min)
72°C for 2 min
35 high stringency cycles:
94°C for 1 min and 30 s
56°C for 1 min and 30 s
72°C for 1 min and 30 s
72°C for 5 min
4°C

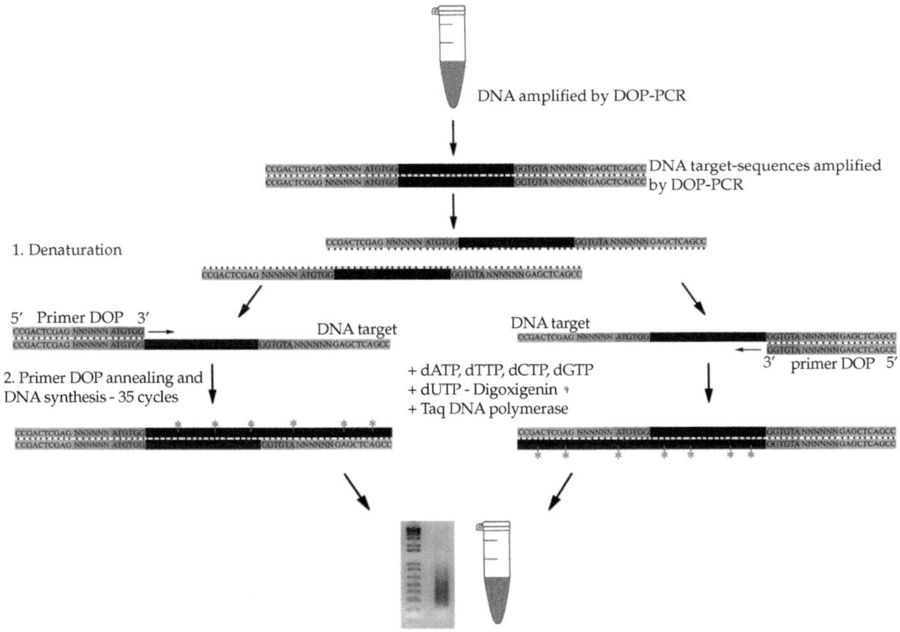

Figure 3. Basic scheme for labeling of microdissected chromosomes by DOP-PCR and digoxigenin dUTP.

Eletrophoresis on agarose gel: Prepare a 1% agarose gel solution and apply 2 μL of DOP-PCR reaction. Run the electrophoresis process at 90 volts for 30 min. The expected result is a smear between 100–600 bp with greater intensity at 350 bp.

Reaction—Second DOP-PCR

The main function of the second DOP-PCR reaction is to generate a larger amount of amplified product and increase the number of possible chromosome paintings without exhausting the product of the first DOP-PCR. Prepare the following PCR mix in laminar flow cabinet:

- 100 ng of template DNA (first DOP-PCR product);
- 1x Fideli Taq reaction buffer;
- 2 mM $MgCl_2$;
- 40 M dNTPs,
- 2 μM DOP primer;
- 1 U Fideli Taq DNA polymerase;
- add sterile water to a final volume of 50 μL.

Place the reaction in the thermocycler at:

94°C for 1 min and 30 s
35 cycles:
94°C for 1 min and 30 s
56°C for 1 min and 30 s
72°C for 1 min and 30 s
72°C for 5 min
4°C

Checking the amplification on agarose gel: Prepare a 1% agarose gel solution and perform electrophoresis with 4 μL of the amplification reaction. Utilize a 100 bp molecular weight marker. The expected result is an amplification of DNA segments with sizes ranging from 100 to 500 bp.

DOP-PCR labeling reaction

The role of this DOP-PCR *labeling* is to prepare the WCP probes. The chromosome probes can be indirectly labeled with biotin or digoxigenin or directly with nucleotides bound to fluorophores. This choice is at the researcher's discretion and depends on the conditions of reagents and equipment. In the laminar flow cabinet, prepare the following PCR mix:

For example, indirectly labeled with digoxigenin:

- 100 ng of template DNA (second DOP-PCR product);
- 1x reaction buffer;
- 2 mM $MgCl_2$;
- 40 μM dATP, dGTP and dCTP, 28 μM dTTP, 12 μM 11-dUTP-digoxigenin* (Roche Applied Science);
- 2 μM DOP primer;
- 1 U Taq DNA polymerase.

* The choice of the modified nucleotide can vary and the concentration must comply with the manufacturer's requirements.

Place the mixture of DOP-PCR labeling in the thermocycler and set a cycle equal to that used in the second DOP-PCR reaction. Run the electrophores is on 1% agarose gel. The expected result is a smear between 100–500 bp.

3. Chromosome Painting by FISH

Main steps illustrating the procedure are shown in Fig. 4.

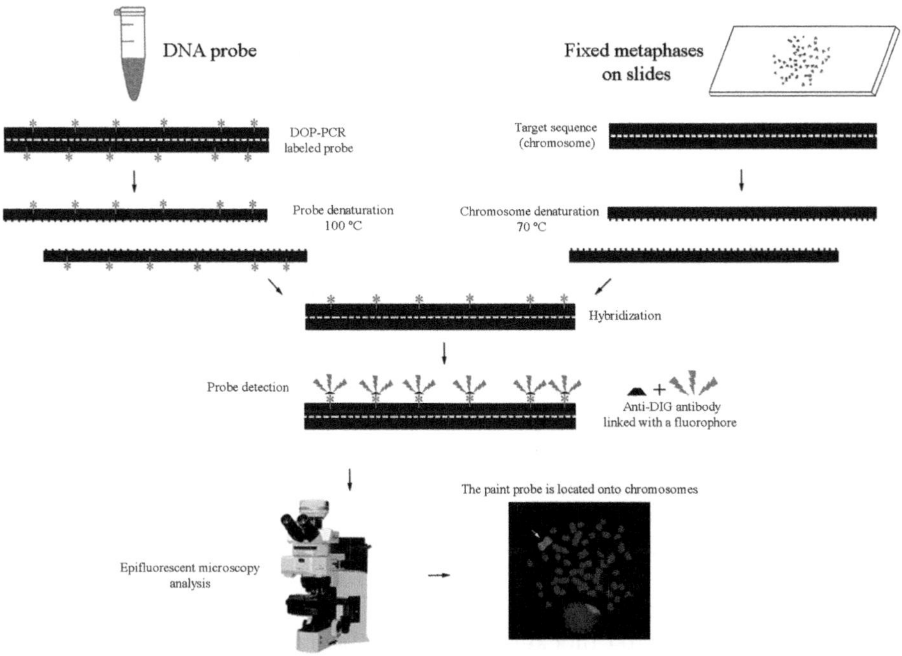

Figure 4. Scheme of chromosome painting with probe obtained from microdissected chromosomes.

Color image of this figure appears in the color plate section at the end of the book.

Step 1: treatment with RNAse

1. Rinse the slides with 1x PBS for 5 min at ambient temperature;
2. Dehydrate the slides in graded series of ethanol (70, 85 and 100%) for 5 min each and allow them to dry;
3. Incubate the slides in 100 μL RNAse (0.4% RNAse/2x SSC) at 37°C for 1 hr in a moist chamber with milli-Q water;
4. Rinse 3 times with 2x SSC for 5 min;
5. Rinse with 1x PBS for 5 min.

Note: Prepare the pepsin and the formaldehyde in RNAse during the incubation. The FISH slide can only dry after dehydration in the ethanol series.

Step 2: treatment with Pepsin

1. Incubate the slides in a 0.005% pepsin solution (in 10 mM HCl) at 37°C for 10 min;
2. Rinse the slides with 1x PBS for 5 min (shaker) at ambient temperature.

Step 3: fixation

1. Fix in 1% formaldehyde 1% in PBS 1x/50 mM $MgCl_2$ for 10 min at ambient temperature;
2. Rinse with 1x PBS for 5 min (shaker);
3. Dehydrate the slides in ethanol series (70, 85 and 100%) for 5 min each.

Note: During this stage, prepare and heat the formaldehyde solution to be utilized in the next step.

Step 4: prehybridization

1. Synthesize the hybridization solution containing 50% formamide, 10% dextransulfate, 2x SSC, 200–300 ηg probe labeled for each slide, 400–600 ηg C_0t-1* blocker for each slide. Use the following model (50 µL hybridization mix) sufficient for 1 slide.
   Hybridization mix (stringency around 77%)
   - 25 µL formamide (50% formamide);
   - 10 µL dextran Suphate 50% (final concentration 10%);
   - 5 µL of 20x SSC (final concentration 2x SSC);
   - x µL of C_0t-1 blocker (volume sufficient for a total of 400–600 ηg);
   - x µL of probe (volume sufficient for a total of 200–300 ηg);
   - Final volume: 50 µL.

   * The solution of C_0t-1 blocker may be synthesized from the specific DNA of the target organism (Zwick et al. 1997).
2. Denature the hybridization solution at 100°C for 10 min and subsequently allow for 30 min prehybridization at 37°C with the blocking probe.
3. Denature the chromosomal DNA with 70% formamide in 2x SSC at 70°C for 5 min.
4. Dehydrate the material in a series of 70, 85 and 100 v/v ethanol for 5 min each.

Note: The ethanol series must be at –20°C.

Step 5: hybridization

1. Prepare the moist chamber at 37°C;
2. Mount the slide with 50 µL hybridization solution, cover with a coverslip and incubate at 37°C overnight or at intervals between 16–48 hr;

Step 6: stringency washes

1. Rinse 2 times in formamide 15%/30% 0.2x SSC pH 7.0 at 42°C for 10 min each;
2. Rinse in a solution of Tween 0.5%/4x SSC, for 5 min at ambient temperature (shaker).

Step 7 a

Note: In the event of direct labeling of the probe.

1. Dehydrate the slides in ethanol series (70, 85, 100 v/v) for 5 min each;
2. Counter stain the slide with 25 µL of DAPI/antifade solution (0.2 µg/mL).

Step 7 b

Note: Only for probes with indirect labeling.
Blocking

1. Incubate the slides in 5% non-fat dry milk—NFDM/4x SSC for 15 min.

Note: Reserve an amount of the 5% NFDM/4x SSC sufficient for mounting the detection antibodies.

2. Rinse twice in Tween 0.5%/4x SSC for 5 min at ambient temperature (shaker).

Step 8: detection of Digoxigenin-labeled probes

1. Incubate the slides in 100 µL of 1 mM anti-Digoxigenin rhodamine conjugated with a 1 : 200 dilution in NFDM/4x SSC buffer (0.5 µL anti-Digoxigenin rhodamine in 99.5 NFDM/4x SSC) in a dark humid chamber for 1 hr at room temperature;
2. Rinse 3 times in Tween 0.5%/4x SSC for 5 min at ambient temperature (shaker);
3. Dehydrate in a series of ethanol solution of 70, 85 and 100 v/v for 5 min each (allow them to dry).

Note: In the case of Biotin-labeled probes, use antibodies conjugated with specific fluorophore.

Step 9: counterstaining

1. Counterstain the slide with 25 µL of DAPI/antifade solution (0.2 µg/mL).

Troubleshooting

Chromosome microdissection

Difficulty in identifying the precise target to be microdissected: old, poorly fixed chromosome preparations with an excessive cytoplasmic content shall not be used for microdissection procedures. Similarly, chromosome preparations with poorly defined morphologies greatly hinder the procedure. In such cases, the best alternative is to make fresh, good quality chromosome preparations.

Excess cytoplasm and membranes hinder microdissection: Meiotic preparations generally retain excess cytoplasm and membranes. For removal of this cytoplasmic material, perform baths with PBS, 0.005%/10 mM HCl pepsin and 4% formaldehyde.

Contamination: A human cell has dozens of times more DNA than a microdissected target chromosome. Thus, bacteria and fungi in reagents and ustensils should be thoroughly checked. We have found that ultraviolet (UV) effects alone are not sufficient to prevent chromosome contamination during the microdissection tests, as they only fragment the DNA. In this case, the material should be autoclaved and the researcher should take basic care in handling, such as wearing gloves and mask and manipulating reagents inside the laminar flow biosafety cabinet.

Whole genome amplification from microdissected products

There was amplification in negative control: the DOP-PCR amplification method involves several handling steps. If the negative control presented amplification on 1% agarose gel (be careful: do not confuse this with the rest of the primer, usually less than 50 bp) some alternatives should be tested. Repeat the procedure with fresh reagents. Check for the sterilization of nozzles, tubes, etc. Check for contamination of primers, nucleotides, buffer, and water used in the reaction. Use tips with filters.

There was no sample amplification: test enzymes and reagents; verify whether the initial amount of microdissected chromosome was not too small; determine whether the chromosome denaturation before addition of the Thermo Sequenase was properly performed; check whether the ramp temperature during the nonspecific amplification cycle of the first DOP-PCR is correct.

The amplification product of the first DOP-PCR produced a low concentration (typically less than 100 ng/µL) or dragout of the expected size: the first DOP-PCR has a critical, initial amplification stage of nonspecific amplification.

This protocol is designed so that minor alterations to the first DOP-PCR nonspecific cycle can be made, such as some adjustment of the annealing temperature and/or a small adjustment of the ramp temperature. We have noticed that very slow (less than that described) or very fast ramps substantially decrease the amplification. Another possibility for the inaccurate amplification is the use of a small amount of the initial DNA. In this case, increase the number of microdissected chromosomes.

Chromosome painting by FISH

Absence of hybridization signal and background: the process of probe synthesis and labeling failed. Relabel the probe.

Excess hybridization signal in several chromosomes hindering the target recognition: Increase the concentration of the blocker and the time of pre-hybridization and suppression.

The hybridization signal is very weak: very weak hybridization signals may result from different situations. (1) The probe may have low incorporation of labeled nucleotides. In this case, a new process for labeling the probe should be conducted. (2) Check the temperature of hybridization and post-hybridization stringency baths. Higher temperatures increase stringency, which can substantially reduce the signal. (3) Verify the concentration of reagents employed in the hybridization mix synthesis. The stringency of the hybridization mix is designed for approximately 77%. Any change in the concentration will generate a variation of the stringency used (Schwarzacher and Heslop-Harrison 2000). (4) Check the concentration of the probe used in the hybridization mix. A total of 100 ηg of labeled probe per slide should be sufficient to produce a good hybridization signal. Concentrations below this should not be used. (5) Excess blocker. The use of an excessive concentration of blocker can substantially reduce the hybridization signal. In this case, adjust the concentrations.

The hybridization occurs, but there is the formation of a halo background hampering visualization: poorly fixed chromosome preparations generally present excess cytoplasm. In this case, best fixed preparations are the best solution. Also, a more aggressive fixation stage with 4% paraformaldehyde can be tested.

Equipment

Major

Chromosome microdissection: Autoclave; laminar flow (biosafety cabinet); micropipette puller; inverted microscope; mechanical micromanipulator.

Whole genome amplification: laminar flow cabinet; thermal cycler; biological samples quantifier; water bath or thermo block; electrophoresis systems.

Chromosome painting: Fluorescence microscope; hybridization oven; biological sample quantifier; water bath or thermo block.

Minor

Chromosome microdissection: Coplin jars; pasteur pipettes; coverslip 24 x 60 mm; Timer; 1 mm x 0.4 mm capillary; microtubes 0.2 mL; micropiptetes; tips.

Whole genome amplification: 0.2 mL microtubes; micropipettes; tips with filters.

Chromosome painting: microtubes; micropipettes; tips; coverslips; slides; Coplin jars; beaker.

Biochemicals and Reagents

Chromosome microdissection:

Giemsa staining; phosphate buffer pH = 6.8.

Whole genome amplification

dNTPs mix; DOP primer (5′-CCGACTCGAGNNNNNNATGTGG-3′); thermosequenase; thermo sequenase buffer (concentrated); water (nuclease-free); dNTP set (dATP, dCTP, dGTP, dTTP); Digoxigenin 11-dUTP (this modified nucleotide can be altered at the researcher's discretion); FideliTaq Polymerase; FideliTaq buffer; $MgCl_2$; Taq polymerase; agarose.

Chromosome painting

Ethanol 70 v/v; ethanol 85 v/v; ethanol 100 v/v; RNAse 10 ηg/μL; pepsin 0.005%/10 mM; HCl; formaldehyde 1%/PBS 1x/50 mM $MgCl_2$; dextransulphate 50%; formamide 70%/30% 2x SSC; Tween 0.5%/4x SSC; Antifade/DAPI (0.2 μg/mL); 5% non-fat dry milk—NFDM/4x SSC

Stock and working Solutions

Pepsin 0.005%/10 mMHCl
99 mL of H_2O
1 mL of HCl 1 M
50 μL of 10% pepsin

Formamide 70%/30% 2x SSC

70 mL of formamide
30 mL of 2x SSC

Formamide 15%/30% 0.2x SSC

30 mL of formamide
20 mL of 2x SSC
150 mL of distilled H_2O

Dextransulphate – 50%

0.5 g of dextransulphate
Dissolve in 1 mL of distilled water

Tween 0.5%/4x SSC

200 mL of 20x SSC
5 mL stock solution Tween 10% or 500 mL pure Tween
Complete the volume to 1000 mL with H_2O

5% non-fat dry milk—NFDM/4x SSC

20 mL of 20x SSC
80 mL of milli-Q water
5 g non-fat dry milk

NaCl – 5 M

29.22 g of NaCl
Dissolve in 100 mL of distilled water, vacuum filter and sterilize in autoclave

Tris HCl – 1 M

trizma base 121.14 g
H_2O 500 mL
HCl up to the desired pH
H_2O up to 1000 mL
Vacuum filter and sterilize in autoclave

Formaldehyde 1%/PBS 1x/50 mM $MgCl_2$

10 mL of PBS 10x
5 mL of $MgCl_2$ 1 M
1 mL of formaldehyde
Complete the volume to 100 mL with distilled H_2O

RNAse 10 mg/mL

Dissolve 10 mg of pancreatic RNAse (RNAse A) in 1 mL of 10 mMTrisHCl pH 7.5 and 15 mMNaCl
Heat at 100°C for 10 min
Store at –20°C

References

Alkan, C., S. Sajjadian and E.E. Eichler. 2010. Limitations of next-generation genome sequence assembly. Nature Methods 8: 61–65.

Arens, Y.H.J.M., J.J.M. Engelen, L.C.P. Govaerts, C.M. van Ravenswaay, W.H. Loneus, J.C.M. van Lent-Albrechts, M. van der Blij-Phillipsen, A.J.H. Hamers and C.T.R.M. Schrander-Stumpel. 2004. Familial insertion (3,5) (q25.3; q22.1q31.3) with deletion or duplication of chromosome region 5q22.1–5q31.3 in ten unbalanced carriers. Am. J. Med. Genet. 130A: 128–133.

Bannai, M., K. Higuchi, T. Akesaka, M. Furukawa, M. Yamaoka, K. Sato and K. Tokunaga. 2004. Single-nucleotide-polymorphism genotyping for whole-genome-amplified samples using automated fluorescence correlation spectroscopy. Anal. Biochem. 327: 215–221.

Bates, G.P., B.J. Wainwright, R. Williamson and S.D.M. Brown. 1986. Microdissection e microcloning from the short arm of human chromosome. Mol. Cell. Bio. 6: 3826–3830.

Beklemisheva, V.R., S.A. Romanenko, L.S. Biltueva, V.A. Trifonov, N.V. Vorobieva, N.A. Serdukova, N.V. Rubtsova, O.V. Brandler, P.C.M. O'Brien, F. Yang, R. Stanyon, M.A. Ferguson-Smith and A.S. Graphodatsky. 2011. Reconstruction of karyotype evolution in core Glires. I. The genome homology revealed by comparative chromosome painting. Chromosome Res. 19: 549–565.

Biltueva, L., N. Vorobieva, P. Perelman, V.A. Trifonov, V. Volobouev, V. Panov, V. Ilyashenko, S. Onischenko, P. O'Brien, F. Yang, M. Ferguson-Smith and A. Graphodatsky. 2011. Karyotype evolution of Eulipotyphla (Insectivora): the genome homology of seven *Sorex* species revealed by comparative chromosome painting and banding data. Cytogenet. Genome Res. 135: 51–64.

Chambeyron, S. and W.A. Bickmore. 2004. Chromatin decondensation and nuclear reorganization of the HoxBlocus upon induction of transcription. Genes Devel. 18: 1119–1130.

Charlesworth, D., B. Charlesworth and G. Marais. 2005. Steps in the evolution of heteromorphic sex chromosomes. Heredity 95: 118–128.

Chen, N., D.W. Bellott, D.C. Page and A.G. Clark. 2012. Identification of avian W-linked contigs by short-read sequencing. BMC Genomics 13: 183.

Chen, Z., R. Morgan, J.F. Stone and A.A. Sandberg. 1993. FISH: a useful technique in the verification of clonality of random chromosome abnormalities. Cancer Genet. Cytogenet. 66: 73–74.

Chowdhary, B.P. and T. Raudsepp. Chromosome painting in farm, pet and wild animal species. Methods Cell. Sci. 23: 37–55.

Cioffi, M.B., A. Sánchez, J.A. Marchal, N. Kosyakova, T. Liehr, V. Trifonov and L.A.C. Bertollo. 2011a. Whole chromosome painting reveals independent origin of sex chromosomes in closely related forms of a fish species. Genetica 139: 1065–1072.

Cioffi, M.B., A. Sánchez, J.A. Marchal, N. Kosyakova, T. Liehr, V. Trifonov and L.A.C. Bertollo. 2011b. Cross-species chromosome painting tracks the independent origin of multiple sex chromosomes in two cofamiliar Erythrinidae fishes. BMC Evol. Biol. 11: 186.

Diniz, D., A. Laudicina, M.B. Cioffi and L.A.C. Bertollo. 2008. Microdissection and whole chromosome painting. Improving sex chromosome analysis in *Triportheus* (Teleostei, Characiformes). Cytogenet. Gen. Res. 122: 163–168.

Falzetti, D., J.R. Vermeesch, C. Matteucci, S. Ciolli, M.F. Martelli, P. Marynen and C. Mecucci. 2000. Microdissection and FISH investigations in acute myeloid leukemia: a step forward to full identification of complex karyotypic changes. Cancer Genet. Cytogenet. 118: 28–34.

Fauth, C. and M.R. Speicher. 2001. Classifying by colors: FISH-based genome analysis. Cytogenet. Cell Genet. 93: 1–10.

Ferguson-Smith, M.A. and V. Trifonov. 2007. Mammalian karyotype evolution. Nature 8: 950–962.

Gracia, E., M.E. Ray, M.H. Polymeropoulos, A. Dehejia, P.S. Meltzer and J.M. Trent. 1997. Isolation of chromosome-specific ESTs by microdissection-mediated cDNA capture. Genome Res. 7: 100–107.

Green, D.M. 1990. Muller's Ratchet and the evolution of supernumerary chromosomes. Genome 33: 818–824.

Green, E.D. 2001. Strategies for the systematic sequencing of complex genomes. Nature Rev. Gen. 2: 573.

Gribble, S., B.L. Ng, E. Prigmore, D.C. Burford and N.P. Carter. 2004. Chromosome paints from single copies of chromosomes. Chromosome Res. 12: 143–151.

Guan, X.Y., P.S. Meltzer, J. Cao and J.M. Trent. 1992. Rapid generation of region specific genomic clones by chromosome microdissection: isolation of DNA from a region frequently deleted in malignant melanoma. Genomics 14: 680–684.

Henning, F., C.B. Moysés, D. Calcagnotto, A. Meyer and L.F. Almeida-Toledo. 2011. Independent fusions and recent origins of sex chromosomes in the evolution and diversification of glass knife fishes (*Eigenmannia*). Heredity 106: 391–400.

Hirai, H., Y. Hirai and P.T. LoVerde. 2012. Evolution of sex chromosomes ZW of *Schistosoma mansoni* inferred from chromosome paint and BAC mapping analyses. Parasitol. Int. http://dx.doi.org/10.1016/j.parint.2012.07.007.

Höckner, M., M. Erdel, A. Spreiz, G. Utermann and D. Kotzot. 2009. Whole genome amplification from microdissected chromosomes. Cytogenet. Genome Res. 125: 98–102.

Kim, D.G., S.Y. Park, K.R. You, G.B. Lee, H. Kim, W.S. Moon, Y.H. Chun and S.H. Park. 2001. Establishment and characterization of chromosomal aberrations in human cholangiocarcinoma cell lines by cross-species color banding. Genes Chromosom. Cancer. 30: 48–56.

Kulemzina, A.I., Y. Fengtang, V.A. Trifonov, O.A. Ryder, M.A. Ferguson-Smith and A.S. Graphodatsky. 2011. Chromosome painting in Tragulidae facilitates the reconstruction of Ruminantia ancestral karyotype. Chromosome Res. 19: 531–539.

Lai, C.S.L., S.E. Fisher, J.A. Hurst, E.R. Levy, S. Hodgson, M. Fox, S. Jeremiah, S. Povey, D.C. Jamison, E.D. Green, F. Vargha-Khadem and A.P. Monaco. 2000. The SPCH1 region on human 7q31: genomic characterization of the critical interval and localization of translocations associated with speech and language disorder. Am. J. Hum. Genet. 67: 357–368.

Langer, P.R., A.A. Waldrop and D.C. Ward. 1981. Enzymatic synthesis of biotin-labeled polynucleotides: Novel nucleic acid affinity probes. Proc. Natl. Acad. Sci. 78: 6633–6637.

Langer, S., C. Fauth, M. Rocchi, J. Murken and M.R. Speicher. 2001. AcroM fluorescent *in situ* hybridization analyses of marker chromosomes. Hum. Genet. 109: 152–158.

Lecerf, F., L. Foggia, P. Mulsant, A. Bonnet and F. Hatey. 2001. A novel method to isolate the common fraction of two DNA samples: hybrid specific amplification (HSA). Nucl. Acids Res. 29(17): e87.

Lengauer, C., M. Speicher, S. Popp, A. Jauch, M. Taniwaki, R. Nagaraja, H.C. Riethman, H. Donis-Keller, M. D'Urso, D. Schlessinger and T. Cremer. 1993. Chromosomal barcodes produced by multicolor fluorescence *in situ* hybridization with multiple YAC clones and whole chromosome painting probes. Hum. Mol. Genet. 2(5): 505–512.

Lichter, P., T. Cremer, J. Borden, L. Manuelidis and D.C. Ward. 1988. Delineation of individual human chromosomes in metaphase and interphase cells by *in situ* suppression hybridization using recombinant DNA libraries. Hum. Genet. 80(3): 224–234.

Lovmar, L., M. Fredriksson, U. Liljedahl, S. Sigurdsson and A.C. Syvanen. 2003. Quantitative evaluation by mini sequencing and microarrays reveals accurate multiplexed SNP genotyping of whole genome amplified DNA. Nucl. Acids Res. 31(21): e129.

Machado, T.C., J.C. Pansonato-Alves, M.B. Pucci, V. Nogaroto, M.C. Almeida, C. Oliveira, F. Foresti, L.C. Bertollo, O. Moreira-Filho, R.F. Artoni and M.R. Vicari. 2011. Chromosomal painting and ZW sex chromosomes differentiation in *Characidium* (Characiformes, Crenuchidae). BMC Genet. 12: 65.

Mahjoubi, F., R.J. Hill and G.B. Peters. 2006. Chromosome microdissection identifies genomic amplifications associated with drug resistance in a leucemia cell line: an approach to understanding drug resistance in cancer. Chromosome Res. 14: 263–276.

Mank, J.E., D.E.L. Promislow and J.C. Avise. 2006. Evolution of alternative sex-determining mechanisms in teleost fishes. Biol. J. Linn. Soc. 87: 83–93.

Mardis, E.R. 2008. Next-generation DNA sequencing methods. Annu. Rev. Genomics Hum. Genet. 9: 387–402.

Metzker, M.L. 2010. Sequencing technologies—the next generation. Nature Rev. Genet. 11: 31.

Nagamachi, C.Y., J.C. Pieczarka, S.S.R. Milhomem, P.C.M. O'Brien, A.C.P. de Souza and M.A. Ferguson-Smith. 2010. Multiple rearrangements in cryptic species of electric knifefish, *Gymnotus carapo* (Gymnotidae, Gymnotiformes) revealed by chromosome painting. BMC Genet. 11: 28.

Nanda, I., P. Benisch, D. Fetting, T. Haaf and M. Schmid. 2011. Synteny conservation of chicken macrochromosomes 1–10 in different avian lineages revealed by cross-species chromosome painting. Cytogenet. Genome Res. 132: 165–181.

Pazian, M.F., C.K. Shimabukuro-Dias, J.C. Pansonato-Alves, C. Oliveira and F. Foresti. 2013. Chromosome painting of Z and W sex chromosomes in *Characidium* (Characiformes, Crenuchidae). Genetica. Online first.

Peng, D.F., H. Sugihara, K. Mukaisho, Y. Tsubosa and T. Hattori. 2003. Alterations of chromosomal copy number during progression of diffuse-type gastric carcinomas: Metaphase- and arraybased comparative genomic hybridization analyses of multiple samples from individual tumours. J. Pathol. 201: 439–450.

Picone, B., J. Masters, D. Silvestro, L. Sineo and M. DelPero. 2011. A phylogenetic analysis of human syntenies revealed by chromosome painting in Euarchontoglires orders. J. Mammal Evol. 18: 131–146.

Pinkel, D., T. Straume and J.W. Gray. 1986. Cytogenetic analysis using quantitative, high-sensitivity, fluorescence hybridization. Proc. Natl. Acad. Sci. 83: 2934–2938.

Ráb, P., M. Rabová, C.S. Pereira, M.J. Collares-Pereira and S. Pelikánová. 2008. Chromosome studies of European cyprinid fishes: interspecific homology of leuciscine cytotaxonomic marker—the largest subtelocêntrico chromosome pair as revealed by cross-species painting. Chromosome Res. 16(6): 863–873.

Rauch, A., R.A. Pfeiffer, U. Trautmann, T. Liehr, H.D. Rott and R.A. Ulmer. 1992. Study of ten small supernumerary (marker) chromosomes identified by fluorescence *in situ* hybridization (FISH). Clin. Genet. 42: 84–90.

Rens, W., P.C.M. O'Brien, F. Grützner, O. Clarke, D. Graphodatskaya, E. Tsend-Ayush, V.A. Trifonov, H. Skelton, M.C. Wallis, S. Johnston, F. Veyrunes, J.A.M. Graves and M.A. Ferguson-Smith. 2007. The multiple sex chromosomes of platypus and echidna are not completely identical and several share homology with the avian Z. Genome Biol. 8: R243.

Röhme, D., H. Fox, B. Herrmann, A.M. Frischauf, J.M. Edström, P. Mains, L.M. Silver and H. Lehrach. 1984. Molecular clones of the mouse t complex derived from microdissected metaphase chromosomes. Cell. 36: 783–788.

Rokas, A. and P.W.H. Holland. 2000. Rare genomics change as a tool for phylogenetics. Trends Ecol. Evol. 15: 454–459.

Rook, M.S., S.M. Delach, G. Deyneko, A. Worlock and J.L. Wolfe. 2004. Whole genome amplification of DNA from laser capture-microdissected tissue for high-throughput single nucleotide polymorphism and short tandem repeat genotyping. Am. J. Pathol. 164: 23–33.

Ropiquet, A., A. Hassanin, E. Pagacova, M. Gerbault-Seureau, H. Cernohorska, S. Kubickova C. Bonillo, J. Rubes and T.J. Robinson. 2010. A paradox revealed: karyotype evolution in the four-horned antelope occurs by tandem fusion (Mammalia, Bovidae, *Tetracerus quadricornis*). Chromosome Res. 18: 277–286.

Scalenghe, F., E. Turco, J.E. Edstrom, V. Pirrotta and M. Melli. 1981. Microdissection and cloning of DNA from a specific region of *Drosophila melanogaster* polytene chromosomes. Chromosoma. 82: 205–216.

Schemberger, M.O., E. Bellafronte, V. Nogaroto, M.C. Almeida, G.S. Schuhli, R.F. Artoni, O. Moreira-Filho and M.R. Vicari. 2011. Differentiation of repetitive DNA sites and sex chromosome systems reveal closely related group in Parodontidae (Actinopterygii: Characiformes). Genetica. 139: 1499–1508.

Schwarzacher, T. and P. Heslop-Harrison. 2000. Practical *in situ* Hybridization. Bios, Oxford. UK.

Telenius, H., N.P. Carter, C.E. Bebb, M. Nordenskjold, B.A. Ponder and A. Tunnacliffe. 1992. Degenerate oligonucleotide-primed PCR: general amplification of target DNA by a single degenerate primer. Genomics. 13: 718–725.

Teruel, M., J. Cabrero, E.E. Montiel, M.J. Acosta, A. Sánchez and J.P.M. Camacho. 2009. Microdissection and chromosome painting of X and B chromosomes in *Locusta migratoria*. Chromosome Res. 17: 11–18.

Traut, W., H. Vogel, G. Glöckner, E. Hartmann and D.G. Heckel. 2013. High-throughput sequencing of a single chromosome: a moth W chromosome. Chromosome Res. Online first.

Vermeesch, J.R., C. Melotte, I. Salden, M. Riegel, V.A. Trifonov, A. Polityko, N. Rumyantseva, I. Naumchik, H. Starke, G. Matthijs, A. Schinzel, J.P. Fryns and T. Liehr. 2005. Tetrasomy 12pter-12p13.31 in a girl with partial Pallister-Killian syndrome phenotype. Eur. J. Hum. Genet. 48: 319–327.

Vicari, M.R., V. Nogaroto, R.B. Noleto, M.M. Cestari, M.B. Cioffi, M.C. Almeida, O. Moreira-Filho, L.A.C. Bertollo and R.F. Artoni. 2010. Satellite DNA and chromosomes in Neotropical fishes: methods, applications and perspectives. J. Fish Biol. 76: 1094–1116.

Vicari, M.R., H.F.M. Pistune, J.P. Castro, M.C. Almeida, L.A.C. Bertollo, O. Moreira-Filho, J.P.M. Camacho and R.F. Artoni. 2011. New insights on the origin of B chromosomes in *Astyanaxscabripinnis* obtained by chromosome painting and FISH. Genetica. 139: 1073–1081.

Volpi, E.V., E. Chevret, T. Jones, R. Vatcheva, J. Williamson, S. Beck, R.D. Campbell, M. Goldsworthy, S.H. Powis, J. Ragoussis, J. Trowsdale and D. Sheer. 2000. Large-scale chromatin organization of the major histocompatibility complex and other regions of human chromosome 6 and its response to interferon in interphase nuclei. J. Cell Science. 113: 1565–1576.

Voltolin, T.A., A. Laudicina, J.A. Senhorini, O. Oliveira, F. Foresti, J. Bortolozzi and F. Porto-Foresti. 2010. Origin and molecular organization of supernumerary chromosome of *Prochilodus lineatus* (Characiformes, Prochilodontidae) obtained by DNA probes. Genetica. 138: 1133–1139.

Weise, A., B. Timmermann, M. Grabherr, M. Werber, P. Heyn, N. Kosyakova, T. Liehr, H. Neitzel, K. Konrat, C. Bommer, C. Dietrich, A. Rajab, R. Reinhardt, S. Mundlos, T.H. Lindner and K. Hoffmann. 2010. High-throughput sequencing of microdissected chromosomal regions. Eur. J.H. Genet. 18: 457–462.

Yoshida, K., Y. Terai, S. Mizoiri, M. Aibara, H. Nishihara, M. Watanabe, A. Kuroiwa, H. Hirai, Y. Hirai, Y. Matsuda and N. Okada. 2011. B chromosomes have a functional effect on female sex determination in Lake Victoria Cichlid fishes. PLoS Genet. 7: e1002203.

Zhang, L., X. Cui, K. Schmitt, R. Hubert, W. Navidi and N. Arnheim. 1992. Whole genome amplification from a single cell: implications for genetic analysis. Proc. Natl. Acad. Sci. 89: 5847–5851.

Zhou, R.N. and Z.M. Hu. 2007. The development of chromosome microdissection and microcloning technique and its applications in genomic research. Curr. Genomics 8(1): 67–72.

Zwick, M.S., R.E. Hanson, M.N. Islam-Faridi, D.M. Stelly, R.A. Wing, H.J. Price and T.D. Mcknight. 1997. A rapid procedure for the isolation of C_ot-1 DNA from plants. Genome 40: 138–142.

19

Extending DNA Fibers and Fiber-FISH

C. Ozouf-Costaz,[1,a,*] *J.P. Coutanceau,*[1,b] *A. D'Hont,*[2] *D. Chassoux*[3] and *G. Lecointre*[4]

[1]Institut de Biologie Paris-Seine, UMR "Evolution Paris-Seine", Case 5, 7 quai Saint Bernard, 75252 Paris Cedex 05.
[a]e-mail: ozouf@mnhn.fr
[b]e-mail: coutance@mnhn.fr
[2]Cirad, UMR AGAP Avenue Agropolis, 34398 Montpellier cedex 5, France.
e-mail: dhont@cirad.fr
[3]Inserm U565, MNHN, 43 rue Cuvier, 75005 Paris.
e-mail: dchassou@mnhn.fr
[4]Institut de Systématique, Evolution et Biodiversité, UMR 7205 CNRS-MNHN-UPMC-EPHE, CP 50, Muséum national d'Histoire naturelle, 57 rue Cuvier, 75005 Paris.
e-mail: lecointr@mnhn.fr
*Corresponding author

Principles

We describe here a protocol of manual stretching (Fig. 1), modified from Jackson (1998) and from Michalet et al. (1977), which uses fish erythrocyte nuclei, a technology which had not been reported before. We have tested it on several species of Antarctic teleosts during several campaigns in the Southern Ocean (ICEFISH, ICOTA-IPEV, REVOLTA-IPEV) and on some eel (*Anguilla anguilla*) specimens, provided by the BOREA laboratory at MNHN. Since the first step requires a cell suspension, corresponding to a precise nuclear DNA concentration, we used erythrocytes, which are nucleated in teleosts and chondrichthyans, and easy to count with a hemocytometer. Briefly, freshly sampled erythrocytes are diluted in PBS and mixed in a buffer containing SDS, to denature proteins, dropped onto a Poly-L-Lysine coated slide and dragged

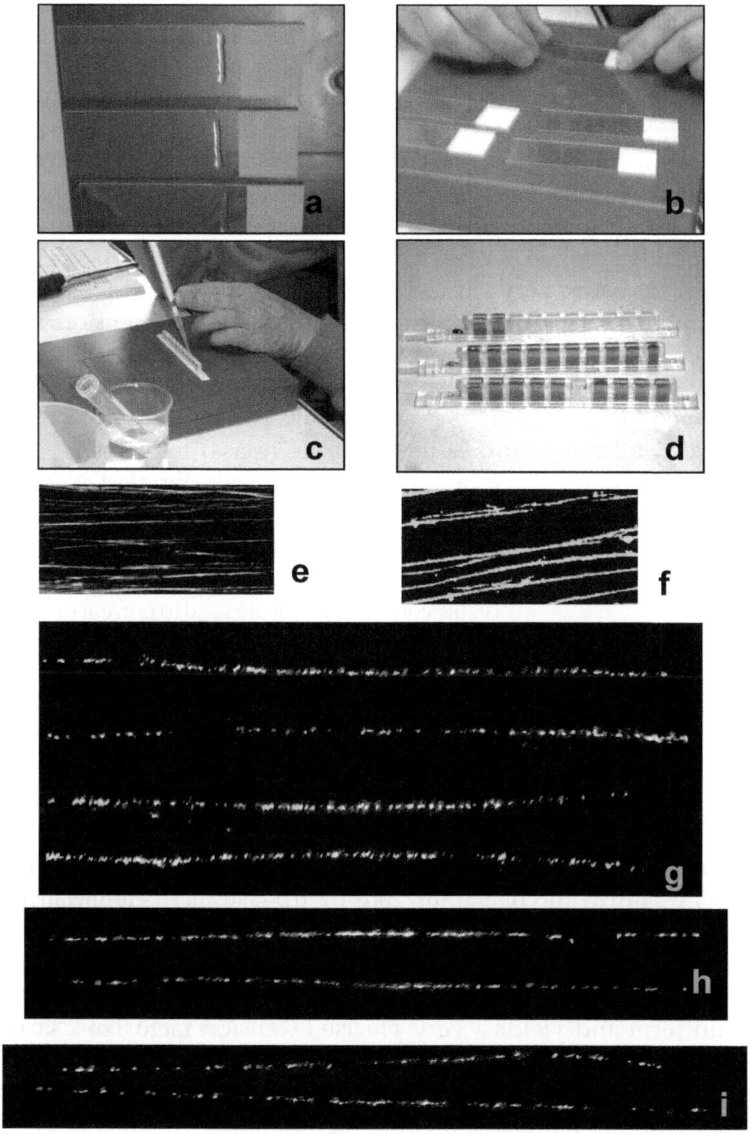

Figure 1. Lines of suspension, along with the frosted part of poly-L-lysine slides (a); dragging the suspension with a large coverslip (b); filling up plug moulds (c and d); YOYO-stained stretched fibers as seen using a fluorescence microscope magnification X40 and X100 (e and f, respectively); double-FISH of a DIG-labeled 5S RNA ribosomal probe (red) and biotin-labeled 28S RNA ribosomal probe (green) on stretched fibers of Antarctic notothenioid fish, as described in § 4, showing alternate copy patterns: *Trematomus bernacchii* (g); *Chionodraco hamatus* (h) and *Trematomus pennellii* (i). For the hemoglobineless channichthyid fish *Chionodraco hamatus,* which has no erythrocytes, we counted lymphocytes, prepared DNA fibers from a concentrated lymphocyte suspension, and obtained similar results.

Color image of this figure appears in the color plate section at the end of the book.

with a large coverslip. Alternatively, the erythrocytes can be embedded in low melting agarose, in order to prevent the DNA fibers from breaking upon subsequent steps. Plugs are treated with proteinase K to inactivate nucleases and release DNA fibers and then can be kept for several months in EDTA at +4°C (Michalet et al. 1997), before being melted and spread onto poly-L-lysine slides for further experiments. When DNA fibers are homogeneously stretched on a solid substrate, DNA is readily accessible to probes and detection reagents. FISH (fluorescence *in situ* hybridization) using various kinds of probes (repetitive cloned probes, single copy gene probes, BAC clones, etc.) is then possible using very similar protocols to those used for chromosome preparations, with slight modifications. However, detection requires a fluorescence microscope equipped with a high sensitivity CCD camera.

Proteinase K treated plugs can also be used for mechanical DNA combing (not described here). It consists of immobilization of high molecular weight DNA molecules (fibres) on a solid substrate (glass) followed by stretching of the molecule by the force of a receding meniscus. Single DNA molecules bind by their extremities onto the surface of glass slides or coverslips coated with silane or poly-L-lysine, and must be uniformly extended and correctly separated and aligned.

Applications and Limitations

Originally developed on human chromosomes to study alterations which do not result in gross chromosomal abnormalities, Fiber-FISH allows researchers to precisely map sequenced probes, to evaluate their size (when they are repeated) and the interval between them, and to understand how they are organized (specially for tandem-repeated sequences, such as ribosomal genes, for instance). DNA can be manually stretched, or mechanically combed (as described in Schurra and Bensimon 2008 for the so-called "dynamic molecular combing"). Manually, the resolution is a few kilobases per micrometer, but with a variable degree of DNA decondensation and fibers can be irregularly stretched, overlapping and not parallel. Mechanically, the DNA stretching is more uniform and yields a very precise extension factor, of 2 kb/µm, but few laboratories have at disposal the suitable equipment and especially silane coated coverslips, which might be difficult to obtain. Information and protocols involving a molecular combing automatic system can be found in the Genomic Vision website (http://www.genomicvision.com/products-instrumentation.php) or in the Leicester University website (http://www.le.ac.uk/bl/phh4/dnacomb.htm), and include many other tips and tricks for such techniques. These are more specifically established for plant and human cell nuclei. For teleost fish, we usually calculate the number of cell nuclei per unit of stretching buffer after prior comparison of the fish genome size with the human genome size.

Protocol

1. Blood sampling

Anaesthetize the fish (see Chapter 1). Using a heparinised syringe, remove a blood sample (at least 0.5 ml and maximum 2 ml) at the fish caudal vein by inserting the needle upwards, posterior to the anal fin, with the bevel facing the fish posterior end. Maintain the syringe in light depression and slowly guide the needle downwards. When it reaches the vein, the blood will enter into the needle hub. Just maintain the depression without sucking, to avoid rupture of the blood vessel. Carefully agitate the blood (do not shake it) in the syringe, to avoid hemolysis or coagulation. Remove the needle and transfer into a sterile tube. The blood sample can be stored at +4°C for up to 24 hours without alteration of the erythrocytes, but should be gently mixed before each use.

2. Estimation of blood cell concentration

Use a hemocytometer (for instance, Malassez hemocytometer) and follow the protocol indicated by the supplier: blood must be diluted in PBS at an appropriate concentration, so that the cells could be well separated from each other in the hemocytometer chamber, with regard to the number of cells to be counted.

3. Calculation of the cell dilution in the PBS buffer prior to stretching (§4) or inclusion in the agarose plug (§5)

The suitable cell concentration for inclusion in agarose, for dynamic molecular combing has been estimated for *Homo sapiens* at 10^7cells/ml (Schurra and Bensimon 2008). We opted for similar DNA concentrations.

To calculate the approximate cell concentration for the teleost fish under study, search in a database (for instance, http://www.genomesize.com/) the genome size (C-value) of the species under study, or alternatively, of its most closely related species, and compare it to the human genome size (3.5 pg).

For instance *Tetraodon nigroviridis* (C-value = 0.35 pg), has a ten times smaller genome than the human genome, its cell concentration [Z] for manual stretching should be thus 10^8 cells/ml.

4. Manual stretching of DNA from live cells *in situ*

This protocol is a modification of Jackson and Pombo (1998) for the cell treatment and of Jackson et al. (1998) for the manual stretching. Main steps of the procedure are described in Fig. 1

Prepare cells at the [Z] concentration in PBS as described in §3.

Drop 20 µl of this suspension on poly-L-lysine coated slide, forming a 1 cm line, along with the frosted part of the slide (Fig. 1a).

Add on the same drop 5 µl of 200 mM Tris-HCl, EDTA 50 mM, 0.5% SDS, pH 7.4.

Keep the slide in horizontal position at laboratory temperature, more or less 5 min, until the edges of the drop start to dry. Then, regularly drag the suspension with a large coverslip maintained tilted, without contacting the surface of the slide (Fig. 1b). At this step, slides can be stored at –20°C, or YOYO-1 stained, to check their quality, or used for FISH experiments.

Note that DNA bound histone proteins may subsist (D. Chassoux, personal communication) and DNA fibers may extend up to varying lengths according to the expected size of duplex DNA of 0.34 nm bp-1 (Parra and Windle 1993). Co-hybridization with probes for closely situated regions of known size may be required for normalization in order to calculate genomic region size. This method is suitable for long range mapping.

5. Manual stretching from agarose embedded cells (from Michalet et al. 1997, modified)

Preparation of agarose plugs

Work with laboratory gloves, and if possible, with sterile materials.

For a 10 well plug mold (100 µl each):

Prepare the blood at the dilution 2 x Z in PBS (for Z concentration, see explanations in §3).

Melt the agarose stock solution in a water bath at 70°C. Then maintain it liquid in another water bath at 45°C.

Indicate species names, sample reference, chosen dilution, date of preparation, etc., on the mould.

In a sterile tube, prepare 500 µl of the 2 x Z diluted blood and mix well by pumping two or three times with a 1 ml Pipetman.

Add 500 µl of liquid agarose, quickly mix to homogenize and very rapidly distribute 95 µl of the mixture in each mould. To avoid bubbles, place the tip of the cone on the side of the plug. It is necessary to work very fast to avoid the agarose solidifying in the pipette (Fig. 1c, d).

Then, put the mould upright in a large Petri dish and let it stay at +4°C between 20 min and maximum two days to let the agarose solidify.

Proteinase K treatment of agarose plugs

Proteinase K buffer solution must be prepared just before use.

Remove the mould from the refrigerator.

Prepare a 50 ml Falcon tube, sterile (one per mould).

Add 5 ml of Proteinase K solution. Note the mould number and details on the tube.

Using the supplier's protocol, let all the plugs fall into the tube.

Close the tube and let it remain overnight and maximum 48 hr in an incubator at 50°C.

Washing the plugs and storing

Remove the tube from the incubator.

Remove supernatant without touching the plugs and fill up the tube with 1 x TE.

Close well and gently mix by slow rotations during 30 min.

At the same time, prepare 1 x TE/1 mM PMSF.

Remove supernatant and replace by 10 ml of 1 x TE/1mM PMSF (1 ml per plug).

Close the tubes and let 30 min in the dark (under aluminium foil) without agitation, on the bench.

Remove supernatant and wash a second time by filling up the tube with 1 x TE during 30 min by slow intermittent rotations.

Reject supernatant.

Replace by filling up the tube with 0.5 M EDTA, pH 8 and store at +4°C for several months (years).

Manual stretching of DNA fibers from agarose embedded cells

The agarose plugs are very fragile and must be manipulated with much care.

Using clean and smooth curved forceps, remove an agarose plug and put it in a round bottom tube containing 10 µl of β-agarase buffer. Heat at 68°C for 15 min and cool down at 40°C (waterbath). After 5 minutes, add 1 µl of β-agarase (1 unit) and leave it at 40°C. After 1 hr, add an additional unit of β-agarase and leave at 40°C for an extra hour (total time of digestion, 2 hr minimum). Add 2 ml of 0.5 M MES, pH 5.5, invert carefully one time and heat at 65°C for 30 min. Invert again carefully and let it cool down progressively at room temperature. Wait 24 hr to let DNA decondense.

Drop 40 µl of this solution on a poly-L-lysine slide, wait about 5 minutes, so that the fibers start to attach and proceed exactly like in § 4 to drag the suspension onto the slide surface.

6. Checking DNA spreading by YOYO-1 staining

YOYO-1 iodide (Molecular Probes, Eugene, OR), is an intercalating dye that stains the DNA backbone and makes it possible for visualization in fluorescence (wavelength exc/em 491/509). It is much brighter than usual counterstainings commonly used for FISH experiments (such as DAPI or propidium iodide).

Dilute the YOYO-1 dye to 2.4 nM in PBS and apply 200 µL of this diluted YOYO-1 solution directly to the slide.

Incubate the sample with the stain at room temperature for 20 minutes.

Briefly rinse with distilled water.

Blot excess of liquid from the slide between two sheets of filter paper.

Mount under coverslip with antifade (see more details in the Chapter 15).

Examine the sample with a fluorescence microscope at magnification x100, equipped with a fluorescein filter set (Fig. 1d).

7. FISH on DNA fibers and signal detection

All kind of probes can be labelled and hybridized on stretched DNA fibers with similar protocols as described in Chapter 15, but they must be concentrated two times more. Just thawed slide preparations must be aged 24 hr at room temperature before denaturation. Stretched DNA must be denaturated at higher temperature than chromosomes, without risk of damage (up to 80°C, 2 minutes) to obtain a more efficient rate of hybridization. Simultanaeous probe-fiber denaturation protocol is thus possible. The most difficult step is to be able to focus on a fiber using the microscope magnification x100. We recommend to previously adjust the focus, for instance, on a DAPI-counterstained chromosome preparation mounted in antifade, then replace by the DNA fiber preparation and change filter. We used a Photometrics coolsnap ES camera and could observe the FISH signals on fibers as described in Michalet et al. 1997, but without amplification with successive layers of antibodies (see examples Fig. 1e). We recommend avoiding these steps because they often produce more background. It is better to use a high sensitivity camera and to amplify the hybridization signals electronically.

Troubleshooting

Optimization of the techniques involves a precise cell concentration in the stretching buffer. Especially for the species for which the genome size could not be estimated precisely, it might be necessary to evaluate the density of fibers on one slide and to increase or decrease the cell concentration in the stretching buffer in the next experiments, in order to obtain the best fiber spreading. In case of doubt, it is better to concentrate the cell suspension more than less. The quality of fiber distribution and stretching can be checked after staining with YOYO-1 DNA stain, but since it requires the of use immersion oil, the slides cannot be used for further FISH experiments. If the DNA is too concentrated, it will clump, or fibers will overlap and look like "ropes". Regularly dragging the slide requires some dexterity and training: if too fast, fibers may break, if too slow, they will not be straight and parallel.

Laboratory set up

Major

pHmeter; fluorescence microscope equipped with x 100 or x 60 objectives and with a high sensitive CCD camera; refrigerator; water baths.

Minor

Syringes and needles; sterile Eppendorf tubes, 2 ml; hemocytometer; micropipettes and cones; poly-L-lysine slides (Fischer Scientific); large coverslips (24 x 50 mm); rounded edge microscope slides; laboratory gloves; 100 µl plug moulds (CHEF disposable plug mould for the preparation of DNA-imbedded agarose plugs, Bio-Rad Laboratories, 2000 Alfred Nobel Drive, Hercules, CA 94547); large Petri dish (or a clean plastic box); Pipetman pipettes; 50 ml sterile tubes.

Biochemicals and Reagents

Heparin sodium (Fluka); Tris-HCl (or Tris-Hydrochloride) (Sigma); EDTA, disodium salt 0,5 M (Sigma Aldrich); SDS (Sodium dodecyl sulphate, for electrophoresis, approx 99%) sterile (Sigma-Aldrich); YOYO-1 iodide stain (Molecular probes); Agarose (low melting) (Invitrogen); Proteinase K, lyophilized (Sigma); PMSF 99% (Sigma); β-agarase 1500 units, recombinant (OZYME); MES 0.5 M in distilled water (Fluka): this buffer must be dissolved in distilled water just before use and maintained at +4°C only for a few days since it can be rapidly damaged; Tris-EDTA 100x for molecular biology (Sigma Aldrich); PBS 1X.

Stock and working solutions

Agarose (low melting) solution, 2% in PBS: Prepare 100 ml, mix and melt at 70°C. Keep in a flask at –20°C between experiments and re-melt for use, or store at –20°C in aliquots.

Proteinase K stock solution: Prepare proteinase K at 20 mg/ml in sterile distilled water, aliquot by 0.5 ml in Eppendorf tubes and store at –20°C between experiments.

Proteinase K working solution: This must be prepared just before use. For one mould of 10 plugs:
Distilled water: 1.75 ml
SDS 10%: 0.5 ml

EDTA 0.5 M: 2.5 ml
Proteinase K 20 mg/ml: 0.25 ml.

Acknowledgements

We are very grateful to CEMIM (CEnter for Microscopic and IMaging at MNHN) where the preliminary experiments were performed.

References

Jackson, D.A. and A. Pombo. 1998. Replicon clusters are stable units of chromosome structure. J. Cell Biol. 140: 1285–1295.

Jackson, S.A., L. Wang, H.M. Goodman and J. Jiang. 1998. Application of fiber-FISH in physical mapping *Arabidopsis thaliana*. Genome 41: 566–572.

Michalet, X., R. Ekong, F. Fougerousse, S. Tousseaux, C. Schurra, N. Hornigold, M. van Slegtenhorst, J. Wolfe, S. Povey, J.S. Beckmann and A. Bensimon. 1997. Dynamic molecular combing: stretching the whole human genome for high resolution studies. Science 277: 1518–1523.

Parra, I. and B. Windle. 1993. High resolution visual mapping of stretched DNA by fluorescent hybridization. Nature 5: 17–21.

Schurra, C. and A. Bensimon. 2008. Combing genomic DNA for structural and functional studies. pp. 71–86. *In*: Hancock, R. (ed.). The Nucleus. Volume 2: Chromatin, Tran cription, Envelope, Proteins, Dynamics and Imaging.

20

In vivo Micronucleus Test for Detecting Response of Teleost Fish to Environmental Genotoxic Exposure

V.N. Phan,[a] *V. Gomes,*[b,*] *M.J.A.C.R. Passos*[c] and *A.J. da Silva Rocha*[d]

Instituto Oceanográfico da Universidade de São Paulo, Praça do Oceanográfico, 191, São Paulo (SP), CEP 05508-120, Brazil.
[a]e-mail: phanvn@usp.br
[b]e-mail: vicgomes@usp.br
[c]e-mail: zezemar@usp.br
[d]e-mail: arrocha.usp.br
*Corresponding author

Principles

Micronuclei are formed by the condensation of acentric chromosomal fragments or whole chromosomes not included in the main nuclei following anaphase (Al-Sabti and Metcalfe 1995). They appear as a result of both chromosome breakage and spindle dysfunction. An increase in the frequency of micronucleated cells can arise from chromosomal and genomic damage caused by clastogens or spindle poisons (Mersch et al. 1996). Micronuclei assay presents several advantages for the study of fish. Micronuclei can be quantified using blood smears, as teleost erythrocytes are nucleated. Counting micronuclei in erythrocytes is easier than the analysis of chromosomal aberrations in metaphase plates. Micronucleus assay *in vitro* or *in vivo* techniques have been developed for use with fish. Several tissues were used to examine the frequency of micronucleated cells in fish captured from contaminated sites or in fish exposed to contaminants *in vivo*. Cephalic kidney cells (Rodriguez-Cea et al. 2003), gill cells (Hayashi et al. 1998; Takai et al. 2004; Bolognesi et al. 2006), hepatic tissue cells (Williams and Metcalfe 1992; Rao et al. 1997; Arcand-Hoy and Metcalfe 2000; Pietrapiana et al. 2002; Çavas et al. 2005); fin cells (Arkhipchuk and Garanko 2005) and embryonic cells (Hayashi et al. 1998) have been used in fish micronucleus assay. However erythrocytes are the most commonly used for micronucleus assay in fish to date. This is partly because fish erythrocytes are nucleated and are easy to obtain and easy to spread as a thin film of cells for examination by microscope. The use of erythrocytes dispenses the need for cell separation. Additionally, there is no need to sacrifice the animal to obtain a small blood sample to make a smear. Blood smear staining is simple and the procedures for analysing blood cell micronuclei and associated erythrocyte nuclear abnormality are well established. In addition the high mitotic rate of hematopoietic tissue provides a rapid response to exposure (Bolognesi et al. 2006).

In micronucleus assay with fish erythrocytes, other erythrocyte nuclear abnormalities, such as kidney shaped nuclei, lobed nuclei or segmented nuclei, have been considered to be analogous of micronuclei (Pacheco and Santos 1997; 1998; Ayllon and Garcia-Vazquez 2000; Gravato and Santos 2002) and used as indicators of genotoxic exposition. The use of erythrocytic nuclear abnormalities in aquatic biomonitoring was discussed in detail by Çavas and Ergene-Gözükara (2003).

In this technical note we will describe methods for *in vivo* fish micronucleus assay particularly with reference to the use of erythrocytes.

Applications and Limitations

Fish are highly diverse vertebrates that occupy a wide variety of habitats. Many fish are concentrated in river creeks, estuaries and shallow water, in water columns or on the ocean floor, where major impact from genotoxic substances is expected. Fish are easily maintained in the laboratory and in the field at relatively low cost, and the range of sizes and characteristics have made fish increasingly important as experimental animals in environmental studies. It has been shown that fish are suitable sentinel organisms for monitoring genotoxic pollutants in the aquatic environment (Al-Sabti and Metcalfe 1995; Hayashi et al. 1998; Bolis et al. 2001; Bolognesi et al. 2006) because they are bioconcentrators and are responsive to mutagens, such as environmental pollutants, at low concentrations.

Concern is growing over the aquatic disposal of industrial and urban wastes, mixing thousands of substances which can be genotoxic to fish and other aquatic organisms. Genotoxic pollutants may interact with the DNA of living cells, induce genetic alterations which may lead to mutations (De Flora et al. 1991; Maccubin et al. 1991) and may induce risk for cancer (Tucker and Preston 1996). The latent effects of the mutations may take considerable time to become apparent and may have adverse consequences on ecosystems (Anderson et al. 1994).

Micronucleus assay has widely been employed as a technique for detecting genotoxic effects, both in the form of laboratory assay, to assess the genotoxicity of a number of compounds, and as *in situ* biomonitoring of environmental mutagenesis, because of its sensitivity, reliability and simplicity of application. Micronucleus assay was originally developed for analysis of the mammalian cell system (Heddle 1973; Schmid 1975) and was subsequently modified for fish by Hooftman and de Raat (1982). This modification, known as the "piscine micronucleus test" (Carrasco et al. 1990) has been widely used for the study of chemical contamination in fish (Das and Nanda 1986; Metcalfe 1988; Al-Sabti and Metcalfe 1995) and other aquatic organisms such as sea urchins (Saotome et al. 1999), bivalves (Mersch et al. 1996), oysters (Burgeot et al. 1995), amphibians (Krauter et al. 1987) and whales (Gauthier et al. 1998).

Protocol

Micronucleus assay in experimental conditions

Fish selection

The choice of the animal is an important step in any experimental design. Several details such as local legislation, sampling facilities, handling and maintenance, suitability and sensibility to tests, sex, size, age, sampling location and ecological importance of fish must be taken into consideration (see "Troubleshooting" section). The animals to be used in experiments should be collected from the same location, at the same time and belong to the same size class. The use of a mixture of individuals collected from different locations, at different times and of different sizes may result in different background levels of micronucleus occurrence.

To avoid unnecessary stress on the animals, fish body weight data for short term experiments can be taken when the fish are processed. If fish body weight data before and after the experiments is necessary, the fish should be identified by their own patterns, if possible, or by tagging.

Maintenance of fish in the laboratory

Usually all recently collected fish are kept together under controlled conditions in storage containers for a period of about one week to one month to allow them to recover after being caught and handed. They are then randomly divided into groups to be used in different treatments of the experiments. Data from at least five fish for each treatment is necessary. To ensure that it is feasible to obtain such data, groups of six to eight fish are recommended, if possible. Groups of fish are then transferred to separate aquaria in the laboratory where they are kept for several days in order to allow the fish to acclimate to the controlled conditions of the laboratory.

Genotoxic substance exposure

In the laboratory, fish can be exposed to a substance by injection or by immersion.

Injection. Intraperitoneal injection, which is simple to apply and allows quick distribution of the substance in the body, is most commonly used. The dose must be calculated so that the volume of solution to be injected is not disproportional to the size of the fish. The choice of substance depends on the aim of the exposure. The selection of concentrations for exposure, injection or immersion, must be carefully studied using toxicity data. A preliminary toxicity test to determine the LD_{50} (Median Lethal Dose) of the substance on the species may be necessary as a reference. Concentrations below or equal to 80% of LD_{50} may be used as a pilot test. Some lipophilic substances are not easily soluble in water. A solvent, such as ethanol, acetone or DMSO must be used as a carrier. The choice of carrier depends on the solubility, homogeneity and stability of the test agent in the carrier (Krishna and Hayashi 2000).

Immersion. Immersion is a considerably more natural way to expose fish to a substance. Exposure by immersion requires detailed planning to maintain the constancy of the experiment such as the reposition of the volatile agent and the changing of solutions. Disposal of used and contaminated water must also be considered.

The quality of water used in the experiments must be controlled and monitored. Variation of temperature, salinity, oxygen, pH, etc. may have an effect on the results of the experiments. Sudden variation in temperature may cause heat shock which is considered genotoxic (causing significant increase of micronuclei) to fish (Anitha et al. 2000).

Time of exposure. For exposure by injection, duration ranging from hour(s) to day(s) is commonly used. For exposure by immersion, duration lasts from day(s) to week(s). Published data indicates the period for optimum harvesting of micronuclei after the exposure. This period depends on the substance, fish type, cell cycle and other unknown factors. Usually three concentrations and three durations are used for exposure (to study the variation of micronucleus based on time of exposure, for example, more intervals may be necessary).

Experimental fish groups

An experiment using five fish, three concentrations and three exposure periods, for example, requires five experimental groups for each exposure period: one negative control group, one carrier control group, and three exposed groups, one for each of the three concentrations (Table 1).

Negative and carrier solvent control groups. Fish from the negative control group remain intact and are submitted to immersion in clean water only. Fish from the carrier control group are those which are submitted to immersion in water with the carrier solvent; and fish from the exposed groups are those which are submitted to immersion in water with the carrier solvent and different concentrations of genotoxic agent. The volume of carrier solvent must be the same for all groups except the negative control.

Positive control group. Besides these groups, it is necessary to have one group as a positive control. Fish from this group are exposed to a well-known genotoxic agent at a concentration sufficient for the induction of a moderate formation of micronuclei in the fish of the group. The inclusion of this group is to insure that the experiment is performed correctly in case other exposed groups produce negative results. The total number of fish necessary for this example can be calculated using the following table (Table 1).

Table 1. Distribution of groups and number of fish necessary for a micronucleus test of a single substance with three concentrations and three periods of exposure.

	Negative ctrl.	**Carrier ctrl.**	**Conc. 1**	**Conc. 2**	**Conc. 3**	**Positive ctrl.***
Period 1	5	5	5	5	5	5
Period 2	5	5	5	5	5	5
Period 3	5	5	5	5	5	5
Sub Total	15	15	15	15	15	15

Total 90 fish; * Optional.

In situ monitoring of genotoxic exposure in field conditions

In situ monitoring can be performed by comparing levels of micronuclei in fish collected at the contaminated site(s) with those of fish of the same species collected at locations presumably free of contamination, which are used as controls. Demersal and non-migratory fish may be a good choice for this kind of investigation.

Caging is also used for *in situ* monitoring (Phan et al. 2007). Fish collected from unpolluted water are divided into groups to be caged at the collection location for use as controls, and from contaminated sites for a pre-determined period. Afterwards, the fish are brought to the laboratory, blood is taken for smearing, and levels of micronuclei are compared. Special cages must be designed and constructed taking into consideration the species, numbers of fish to be accommodated in the cage, and local of exposure. Special care must be taken in order to avoid possible loss of data due to damages or predation caused to the fish by natural occurrences or opportunistic predators.

In situ micronucleus data was sometimes reported together with data on the physical and chemical parameters of the site (Buschini et al. 2004). The correlation between micronucleus frequency and concentration of genotoxin in the tissues has also been reported (Ilyinskikh et al. 1998).

Preparation of blood smears for micronucleus assay

Slide washing. Slides may be washed in alcohol, 50:50 alcohol: ether mixture, or in a solution of 10% of Triton X-100. Dirty slides may be treated by immersing in dichromate cleaning solution for 24 hrs, washed in running tap water and rinsed in distilled water then stored in alcohol (95–100%). Slides are cleaned with soft cloth before use.

Blood sampling. Blood samples can be obtained from various sites on the fish. Caudal aorta or cardiac puncture (the drawing of a sample of blood from the *bulbus arteriosus* or Cuvier duct with a syringe and needle) are commonly used. *Bulbus arteriosus*, as suggested by its name, is an onion shaped bulb, filled with blood and located in front of the ventricle in relation to the gill cleft from where it can be easily accessed by a hypodermic syringe and needle. *Bulbus arteriosus* is an elastic bulb which receives blood from the ventricle, regulates its pressure and sends it to the gill. This pressure makes the drawing of blood easy, and as it is highly elastic it prevents the blood from leaking and seals the hole perforated by the needle rapidly when the needle is removed. With practice, it is possible to collect a sufficient volume of blood to prepare about six smears from a fish with a body weight of about 3 g. The size of syringe and needle must be selected based on the size of the fish. A disposable syringe of 0.5 ml including the attached needle is the most useful for fish with a body weight of 5 to 50 g. With practice, there is no need to use anticoagulants such as heparin or EDTA. Small fish can be immobilized by wrapping them in a damp towel while taking their blood. Anaesthesia may be necessary. Several kinds of anaesthesia are available for fish, of which MS 222 is the most widely used. Cardiac punctures also allow blood samples to be taken repeatedly from the same fish. The interval between two consecutive bleedings must allow the fish to recover from stress and blood loss.

Blood samples can also be obtained from the caudal vein (see protocols Chapter 1 and Chapter 19).

Blood smears. Preparing a good quality smear is one of the most important steps in micronucleus assay.

- Keep a clean and grease-free slide on a horizontal surface.
- Immediately after drawing the blood, place a small drop of blood in the middle of the slide, about 2 cm from one end.
- Use a second slide as a spreader.
- Hold the spreader firmly on the surface of the first slide about 1–2 cm from the blood drop.
- Incline the spreader in the direction of the blood drop until the two slides form an angle of about 30 degree.
- Draw the spreader toward the blood until contact is made, whereupon the blood will quickly and evenly spread along the edge of the spreader.

- Push the spreader rapidly, smoothly and evenly in the opposite direction, maintaining the contact and the angle between the two slides.
- Leave the spreader and hold the first slide by its short edges then rapidly move it horizontally to dry the film of blood for about 15 s. This step is important to maintain the shapes of the blood cells.
- Keep the slide on the horizontal surface for further drying for several hours.
- Fix the slides in absolute methanol in a Coplin jar for 3 to 5 min.
- Air dry the slides and keep them in a slide box at room temperature for later staining.
- At least two slides are necessary for each fish.

Thick blood films give overlapping images of the cells and make analysis difficult. Thin films make cells more distant from each other and facilitate the analysis.

The thickness of the film depends on the size of the drop, the viscosity of the blood, the velocity of the spreader (the more rapid the movement, the thinner the film) and the angle at which the spreader is held (the bigger the angle, the thicker the smear).

Staining blood smears

Giemsa staining. The slides fixed with methanol and air-dried are immersed in a diluted Giemsa stain in a Coplin jar for 15 to 30 min, according to the Giemsa batch used. The slides are then washed in tap water, rinsed in distilled water and air dried.

When completely dried, the slides are ready to be examined. The smear is not covered. An example of a micronucleated cell obtained by this procedure is shown in Fig. 1.

Several others staining methods can be used to reveal the micronuclei in fish, such as double Giemsa-May-Grünwald (Rodriguez-Cea et al. 2003), hematoxylin-eosin (Farah et al. 2005), Feulgen (Minissi et al. 1996) and fluorescent dyes such as acridine orange staining (Ueda et al. 1992; Hayashi et al. 1998; Takai et al. 2004; Çavas and Ergene-Gözükara 2005).

Assess the micronucleated cell frequency

Principal characteristics used as criteria for identifying micronuclei are as follows: (1) micronuclei are spherical cytoplasmic inclusions with a sharp contour, (2) diameter ranges between 1/10 and 1/30 of the nucleus, (3) the micronuclei resemble the nuclei in both texture and colour, (4) they have no contact with the nucleus, (5) they are non-refractile particles and are on the same optical plane as the main nucleus.

Micronucleated cell frequency is calculated as the number of micronucleated cells per 1000 cells. To count micronucleated cells, stained and dried slides are

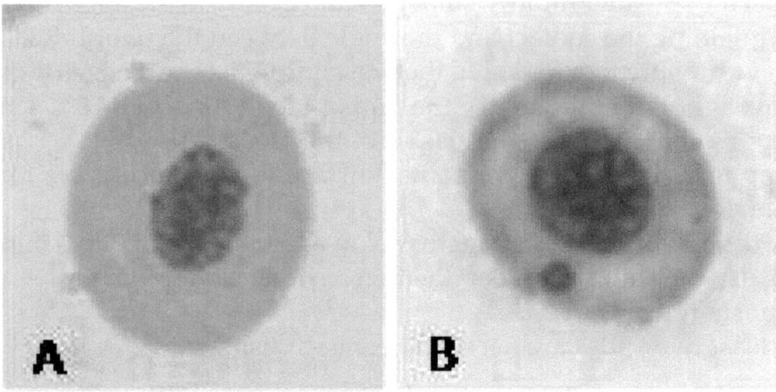

Figure 1. Erythrocytes from Antarctic fish *Trematomus newnesi.* (A). Mature erythrocytes with normal nucleus; (B). Micronucleated cell. Giemsa staining.

coded, randomized and blindly scored by a single trained researcher under a microscope at 1000x magnification. Data of at least one replicate for each fish is recommended.

Bolognesi et al. (2006) discarded cells with more than four micronuclei to exclude apoptotic phenomena. Cells with abnormal morphological forms (degenerated cells) induced by cytotoxic events were also discarded. Minissi et al. (1996) counted only cells with one micronucleus. The sizes of micronuclei used by the authors varied between 1/5 and 1/100 of the main nucleus. To determine the size of micronucleus, it should be noted that the size and the number of chromosomes vary with fish species.

Bolognesi et al. (2006) analysed the baseline micronucleated cell frequency obtained by different authors from existing literature and reported a large variation in frequency (0–13) between different species as well as a large variation in frequency (up to 1–2 orders of magnitude) of the same species obtained by different authors. The authors emphasize the importance of standardization and intercalibration of the micronucleus assay experimental procedure with particular reference to scoring criteria so that data can be compared between laboratories.

Statistical analysis

Multisample analyses are frequently used for the statistical analysis of micronucleus assay data. Several authors first tested data normality and variance homogeneity to meet assumptions of parametric statistical procedures, then applied ANOVA to determine differences among treated groups, followed by Newman Keuls or Tukey post hoc tests to identify differences between groups. Other authors however, applied nonparametric the ANOVA Kruskal-Wallis test to analyse differences among groups, then used the Mann-Whitney U-test to compare differences between groups and between groups and control. Several commercial statistics programs test the supplied data automatically, then guide users through analysis with appropriate statistical procedures. There are also statistical tests that use binominal distribution or Poisson distribution (Kastenbaum and Bowman 1970; Elder 1992) to test the frequency of treated and control groups.

Troubleshooting

Species and locality of sampling

Differences in sensitivity to environmental exposure have been described for freshwater and seawater fish. Pietrapiana et al. (2002) used hepatic micronuclei as an indicator to detect environmental hydrocarbon exposure in three demersal fish species with different habitats and feeding behaviour (i.e., *Lepidorhombus boscii, Merluccius merluccius* and *Mullus barbatus*). The authors reported a difference in sensitivity between the species used. *L. boscii* was more sensitive in terms of both micronuclei frequency and tissue damage. Hayashi et al. (1998) reported a difference in gill cell micronucleus frequencies in *Carassius* sp. (Funa) and *Zacco platypus* (Oikawa) collected from upstream and midstream positions of the Tomio River (Nara, Japan). Fish collected upstream tended to have lower micronucleus frequencies than those collected at the midstream position of the river.

Seasonal differences

Hayashi et al. (1998) reported seasonal differences in *Carassius* sp. (Funa) and *Zacco platypus* (Oikawa) collected from the Tomio River (Nara, Japan), *Leiognathus nuchalis* (Hiiragi) as well as in *Ditrema temmincki* (Umitanago), small marine fish collected periodically from Mochimune Harbor, Shizuoka, Japan. For *Carassius* sp. (Funa) and *Zacco platypus* frequency of micronucleated gill cells and micronucleated erythrocytes are generally higher in May than in June, August and September. For *L. nuchalis* and *D. temmincki* higher frequencies are generally reported in July through September. Minissi et al. (1996), on the other hand, reported no seasonal difference in erythrocyte micronuclei frequencies in *Barbus plebejus* collected from two rivers of central Italy in different months.

Sex, size and age

The sex of immature fish is generally difficult to determine unless the fish is sexually dimorphic. The sex of adult fish in the spawning season can be determined by the shape of the belly or by pressing on the belly to see if eggs or semen come out of the fish. For immature fish, sex is usually determined at the end of the experiment when the fish is sacrificed and the gonad is removed for examination under microscope. If data of the sex of experimental fish is necessary, the number of fish in experimental groups must be adequately increased so that data can be separated later. The body size of fish is an important factor to be considered for laboratory experiments. Small fish make sampling difficult. An increase in the body size of fish means an increase in experiment water volume, agents, space for aquaria and water for disposal. Fish from 20 to 50 grams in body weight are suitable for the first trial, whereas working with fish weighing less than 20 grams may require a little practice. Age determination is another problem. With practice, the age of a fish can be determined by reading year rings on scales, otoliths, and the cross-section of fins or vertebrae.

Takai et al. (2004) studied the susceptibility of male and female medaka (*Oryzias latipes*) to spontaneous and X-ray induced micronuclei in gill cells and concluded that in X-ray treated fish, micronucleated cell frequency and body weight were not significantly correlated, nor was there any difference between sexes. Ilyinskikh et al. (1998) conducted cytogenetic monitoring of *Esox lucius* pikes contaminated by radiocesium in the Tomsk River, Russia, using micronucleus assay, and found that there were no statistically significant differences between the micronucleated erythrocytes in males and females aged from one to ten. However, males aged over 20 yrs were statistically different from females.

Laboratory set up

Equipment

Major

Light microscope with digital camera; fluorescence microscope equipped with camera and suitable filter set; pHmeter.

Minor

Disposable syringes and needles (at least two syringes and needles must be provided for each fish; the sizes of the syringes and needles must be selected in accordance with the size of fish); microscope slides and slide boxes; Coplin jars, beakers and laboratory glass wares.

Biochemicals and reagents

Absolute methanol P.A.; Acetone P.A.; DMSO P.A.; Ether P.A.; Ethanol P.A.; Triton X-100; Anticoagulants such as heparin or EDTA; MS 222 or other anaesthetics.

Stock and working solutions

GIEMSA

1 volume of stain in 9 to 15 volumes of distilled water or phosphate buffer at pH 6.8 freshly prepared before use.

References

Al-Sabti, K. and C.D. Metcalfe. 1995. Genetic toxicology fish micronuclei for assessing genotoxicity in water. Mutat. Res. 343: 121–135.

Anderson, S., W. Sadinski, L. Shugart, P. Brussard, M.H. Depledge, T. Ford, H. Hose, J.J. Stegeman, W. Suk, I. Wirgin and G. Wogan. 1994. Genetic and Molecular Toxicology: A research framework. Environ. Hlth. Perspect. 102: 3–8.

Anitha, B., N. Chandra, P.M. Gopinath and G. Durairaj. 2000. Genotoxicity evaluation of heat shock in gold fish (*Carassius auratus*). Mutat. Res. 469: 1–8.

Arcand-Hoy, L.D. and C.D. Metcalfe. 2000. Hepatic micronuclei in brown bullheads (*Amerius nebulosus*) as a biomarkers for exposure to genotoxic chemicals. J. Great Lakes Res. 26: 408–415.

Arkhipchuk, V.V. and N.N. Garanko. 2005. Using the nucleolar biomarker and the micronucleus test on *in vivo* fish fin cells. Ecotoxicol. Environ. Saf. 62: 42–52.

Ayllon, F. and E. Garcia-Vazquez. 2000. Induction of micronuclei and other nuclear abnormalities in European minnow *Phoxinus phoxinus* and mollie *Poecilia latipinna*: an assessment of the fish micronucleus test. Mutat. Res. 467: 177–186.

Bolis, C.L., M. Piccolella, A.Z. Dalla Valle and J.C. Rankin. 2001. Fish as model in pharmacological and biological research. Pharmacol. Res. 44: 265–280.

Bolognesi, C., E. Perrone, P. Roggieri, D.M. Pampanin and A. Sciutto. 2006. Assessment of micronuclei induction in peripheral erythrocytes of fish exposed to xenobiotics under controlled conditions. Aquat. Toxicol. 78S: S93–S98.

Burgeot, T., E. His and F. Galgani. 1995. The micronucleus assay in *Crassostrea gigas* for the detection of seawater genotoxicity. Mutat. Res. 342: 125–140.

Buschini, A., A. Martino, B. Gustavino, M. Monfrinotti, P. Poli, C. Rossi, M. Santoro, A.J.M. Dörr and M. Rizzoni. 2004. Comet assay and micronucleus test in circulating erythrocytes of *Cyprinus carpio* specimens exposed *in situ* to lake waters treated with disinfectants for potabilization. Mutat. Res. 557: 119–129.

Carrasco, K.R., K.L. Tilbury and M.S. Mayers. 1990. Assessment of the piscine micronuclei test as an *in-situ* biological indicator of chemical contaminants effects. Can. J. Fish Aquatic Sci. 47: 2123–2136.

Çavas, T. and S. Ergene-Gözükara. 2003. Micronuclei, nuclear lesions and interphase silver-stained nucleolar organizer regions (AgNORs) as cyto-genotoxicity indicators in *Oreochromis niloticus* exposed to textile mill effluent. Mutat. Res. 538: 81–91.

Çavas, T. and S. Ergene-Gözükara. 2005. Genotoxicity evaluation of metronidazole using the piscine micronucleus test by acridine orange fluorescent staining. Environ. Toxicol. Pharmacol. 19: 107–111.

Çavas, T., N.N. Garanko and V.V. Arkhipchuk. 2005. Induction of micronuclei and binuclei in blood, gill and liver cells of fishes subchronically exposed to cadmium chloride and copper sulphate. Food Chem. Toxicol. 43: 569–574.

Das, R.K. and N.K. Nanda. 1986. Induction of micronuclei in peripheral erythrocytes of fish *Heteropneustes fossilis* by mitomycin C and paper mill effluent. Mutat. Res. 175: 65–71.

De Flora, S., M. Bagnasco and P. Zanacchi. 1991. Genotoxic, carcinogenic, and teratogenic hazard in the marine environment, with special reference to the Mediterranean Sea. Mutat. Res. 258: 285–320.

Elder, L. 1992. Statistical methods for short-term tests in genetic toxicology: the first fifteen years. Mutat. Res. 277: 11–33.

Farah, M.A., B. Ateeq and W. Ahmad. 2005. Antimutagenic effect of neem leaves extract in freshwater fish, *Channa punctatus* evaluated by cytogenetic tests. Sci. Total Environ. 364: 200–214.
Gauthier, J.M., H. Dubeau and E. Rassart. 1998. Mercury induced micronuclei in skin fibroblasts of beluga whales. Environ. Toxicol. Chem. 17: 2487–2493.
Gravato, C. and M.A. Santos. 2002. β-Naphthoflavone liver EROD and erythrocytic nuclear abnormality induction in juvenile *Dicentrarchus labrax* L. Ecotoxicol. Environ. Saf. 52: 69–74.
Hayashi, M., T. Ueda, K. Uyeno, K. Wada, N. Kinae, K. Saotome, N. Tanaka, A. Takai, Y.F. Sasaki, N. Asano, T. Sofuni and Y. Ojima. 1998. Development of genotoxicity assay systems that use aquatic organisms. Mutat. Res. 399: 125–133.
Heddle, J.A. 1973. A rapid *in vivo* test for chromosomal damage. Mutat. Res. 18: 187–190.
Hooftman, R.N. and W.K. de Raat. 1982. Induction of nuclear anomalies (micronuclei) in peripheral blood erythrocytes of Estern mudminnow *Umbra pygmaea* by ethyl methanesulphonate. Mutat. Res. 104: 147–152.
Ilyinskikh, N.N., E.N. Ilyinskikh and I.N. Ilyinskikh. 1998. Micronucleated erythrocytes frequency and radiocesium bioconcentration in pikes (*Esox lucius*) caught in the Tom River near the nuclear facilities of the Siberian Chemical Complex (Tomsk-7). Mutat. Res. 421: 197–203.
Kastenbaum, M.A. and K.O. Bowman. 1970. Tables for determining the statistical significance mutation frequencies. Mutat. Res. 9: 527–549.
Krauter, P.W., S.L. Anderson and F.L. Harisson. 1987. Radiation-induced micronuclei in peripheral erythrocytes of *Rana catesbeiana*: an aquatic animal model for *in vivo* genotoxicity studies. Environ. Mol. Mutagen. 10: 285–291.
Krishna, G. and M. Hayashi. 2000. *In vivo* rodent micronucleus assay: protocol, conduct and data interpretation. Mutat. Res. 455: 155–166.
Maccubin, A.E., N. Ersing and M.E. Frank. 1991. Mutagenicity of sediment from the Detroit River. J. Great Lakes Res. 17: 314–321.
Mersch, J., M.-N. Beauvais and P. Nagel. 1996. Induction of micronuclei in haemocytes and gill cells of zebra mussels, *Dreissena polymorpha*, exposed to clastogens. Mutat. Res. 371: 47–55.
Metcalfe, C.D. 1988. Induction of micronuclei and nuclear abnormalities in the erythrocytes of mudminnows (*Umbra limi*) and brown bullheads (*Ictalurus nebulosus*). Bull. Environ. Contamin. Toxicol. 40: 489–495.
Minissi, S., E. Ciccotti and M. Rizzoni. 1996. Micronucleus test in erythrocytes of *Barbus plebejus* (Teleostei, Pisces) from two natural environments: a bioassay for the *in situ* detection of mutagens in freshwater. Mutat. Res. 367: 245–251.
Pacheco, M. and M.A. Santos. 1997. Induction of EROD Activity and genotoxic effects by polycyclic aromatic hydrocarbons and resin acids on the juvenile eel (*Anguilla anguilla* L.). Ecotoxicol. Environ. Saf. 38: 252–259.
Pacheco, M. and M.A. Santos. 1998. Induction of liver EROD and erythrocytic nuclear abnormalities by cyclophosphamide and PAHs in *Anguilla anguilla* L. Ecotoxicol. Environ. Saf. 40: 71–76.
Phan, V.N., V. Gomes, M.J.A.C.R. Passos and A.J.S. Rocha. 2007. Biomonitoring of the genotoxic potential (micronucleus and erythrocyte nuclear abnormalities assay) of the Admiralty Bay water surrounding the Brazilian Antarctic Research Station "Comandante Ferraz", King George Island. Polar Biol. 30: 209–217.
Pietrapiana, D., M. Modena, P. Guidetti, C. Falugi and M. Vacchi. 2002. Evaluating the genotoxic damage and hepatic tissue alterations in demersal fish species: a case study in the Ligurian Sea (NW-Mediterranean). Mar. Pollut. Bull. 44: 238–243.
Rao, S.S., T. Neheli, J.H. Carey and V.W. Cairns. 1997. Fish hepatic micronuclei as an indication of exposure to genotoxic environmental contaminants. Environ. Toxicol. Water Qual. 12: 217–222.
Rodriguez-Cea, A., F. Ayllon and E. Garcia-Vazquez. 2003. Micronucleus test in freshwater fish species: an evaluation of its sensitivity for application in field surveys. Ecotoxicol. Environ. Saf. 56: 442–448.
Saotome, K., T. Sofuni and M. Hayashi. 1999. A micronucleus assay in sea urchin embryos. Mutat. Res. 446: 121–127.
Schmid, W. 1975. The micronucleus test. Mutat. Res. 31: 9–15.

Takai, A., N. Kagawa and K. Fujikawa. 2004. Susceptibility of male and female medaka (*Oryzias latipes*) fish to spontaneous and X-ray induced micronucleus formation in gill cells. Mutat. Res. 558: 131–136.

Tucker, J.D. and R.J. Preston. 1996. Chromosome aberrations, micronuclei, aneuploidy, sister chromatid exchanges, and cancer risk assessment. Mutat. Res. 365: 147–159.

Ueda, T., M. Hayashi, Y. Ohtsuka, T. Nakamura, J. Kobayashi and T. Sofuni. 1992. A preliminary study of the micronucleus test by acridine orange fluorescent staining compared with chromosomal aberration test using fish erythropoietic and embryonic cells. Water Sci. Technol. 25: 235–240.

Williams, R.C. and C.D. Metcalfe. 1992. Development of an *in vivo* hepatic micronucleus assays with rainbow trout. Aquat. Toxicol. 23: 193–202.

Abbreviations

AD	DAPI/actinomycin
AT	adenine-thymine
ATP	adenosine triphosphate
BAC	bacterial artificial chromosome
bFGF	basic fibroblast growth factor
bp	base pairs
BrdU	5-bromodeoxyuridine
BSA	bovine serum albumin
CBG	centromeric banding Giemsa
CCD	charge-coupled device
CCD	cooled camera device
CDD-banding	chromomycinA_3/distamycin/DAPI banding
cDNA	complementary DNA
CGH	comparative genomic hybridization
CMA_3	chromomycin A_3
CMF-PBS	10 X calcium- and magnesium-free phosphate buffered saline
ConA	concanavalin A
Cot-1 DNA	DNA enriched for highly and moderately repetitive DNA sequences
C_0t-1 repetitive DNA	Cot-1 DNA used as a competitor
CTAB	cetyl trimethylammonium bromide
CTP	cytidine triphosphate
DA	distamycin A
DAPI	4′,6-diamidino-2-phenylindole
dATP	2′-desoxyadenine 5′-triphosphate
dCTP	2′-desoxycytidine 5′-triphosphate
ddH_2O	double-distilled water
dGTP	2′-desoxyguanosine 5′-triphosphate
DMEM	Dulbecco/Vogt modified Eagle's minimal essential medium
DMSO	dimethyl sulfoxide $(CH_3)_2SO$
dNTP 2′	desoxynucleotide 5′-triphosphate
dNTP(s)	deoxynucleotide triphosphate(s)

D_2O	:	distilled water
DOP	:	degenerate oligonucleotide primer
dTTP	:	2′-desoxythymidine 5′-triphosphate
dUTP	:	deoxyuridine triphosphate
dUTP	:	2′-desoxyuridine 5′-triphosphate
EDTA	:	ethylenediaminetetraacetic acid
EtOH	:	ethanol (only undenatured ethanol can be used)
ex/em	:	excitation/emission
FBS	:	fetal bovine serum
FCS	:	fetal calf serum
FISH	:	fluorescence *in situ* hybridization
FITC	:	fluorescein-5-isothiocyanate
FN	:	fundamental number
g	:	$9{,}81\ m\ .\ s^{-2}$
G-bands	:	chromatid bands after treatment with trypsin, then with Giemsa
GC	:	guanine cytosine
gDNA	:	genomic DNA
GISH	:	genomic *in situ* hybridization
GTP	:	guanosine triphosphate
HEPES	:	(4-(2-hydroxyethyl)-1-piperazineethanesulfonic acid)
ISH	:	*in situ* hybridization
KOac	:	potassium acetate salt (CH3COOK, 1½ H_2O)
L-15	:	L-15 Leibowitz culture medium
LB	:	Luria Bertani broth
LPS	:	Lipopolysaccharide
MDA	:	multiple displacement amplification
MES	:	2-(*N*-morpholino)ethanesulfonic acid
MM	:	mithramycin M
MS 222	:	tricaine methanesulfonate
NFDM	:	non-fat dry milk
NGS	:	normal goat serum
NGS	:	next generation sequencing
NORs	:	nucleolar organizing regions
NT	:	nick translation
OD	:	optical density
PBD	:	phosphate buffer detergent
PBS	:	phosphate buffer saline
PCR	:	polymerase chain reaction
PCR mix	:	reagents needed to run a PCR reaction which does not include polymerase and template DNA
PHA	:	phytohaemagglutinin
PMSF	:	phenylmethylsulfonylfluoride
PWM	:	pokeweed mitogen

Q-bands	:	chromatid bands stained with quinacrine
R-bands	:	"reverse" chromatid bands complementary to G-bands
RGB	:	red, green, blue
RPM (or rpm)	:	rotations per minute
RT	:	room temperature
SC	:	Synaptonemal complex
SDS	:	sodium dodecyl sulphate buffer
SNP	:	single nucleotide polymorphism
SSC	:	saline-sodium citrate buffer
SYCP3	:	component of the axial/lateral element of the synaptonemal complex
TE	:	Tris-EDTA
Tris	:	2-Amino-2-hydroxymethyl-propane-1,3-diol
TTP	:	thymidine triphosphate
UV	:	ultra-violet
WCP	:	whole chromosome painting
WGA	:	whole genome amplification
YAC	:	yeast artificial chromosome

Glossary

Antifade: a solution, usually based on glycerol, reducing photobleaching; thus, preserving signals of fluorescently labelled target molecules.

Antifungal (or antimycotic): fungicide used to prevent mycoses.

Antigen recovery: re-exposure of binding sites for the antibody after paraformaldehyde fixation.

Antimycotic: fungicide used to prevent mycoses.

Aplasia: absence of cell divisions.

Autologous serum: serum of the species under study.

B chromosomes: extra chromosomes found in many eukaryotic genomes in addition to the chromosomes of the standard A complement.

Banding: techniques used in cytogenetics to produce patterns of transversal bands characterizing chromosomal pairs.

Chromatin: the main component of chromosome including the DNA, chromosomal proteins and chromosomal RNA.

Chromosome band: transversal band on chromosome revealed by special staining procedures (banding).

Chromosome microdissection: technique that physically cuts a chromosomal region from a chromosome or an entire chromosome from a metaphase plate spread out onto a microscope slide.

Chromosome painting: FISH technique in which whole chromosome-specific probes are used.

Colchicine: toxic natural product and secondary metabolite, originally extracted from plants of the genus *Colchicum.* It acts as a mitotic spindle poison by interfering with microtubule polymerization and allows to accumulate mitotic stages for karyotyping.

Competitor DNA/Cot-1 DNA: unlabelled, physically or enzymatically sheared (ca. 100 bp) gDNA added to the hybridization reaction to saturate high repetitive sequences in order to avoid unspecific hybridization of the probe.

Cosmid: vector used in genetic engineering to clone genes.

Cot-1 DNA: see competitor DNA.

Depurination: cleavage of the N glycosidic link between purine bases and deoxyribose in DNA.

DNA satellite: DNA containing multiple tandem repeats of a unit represented by a sequence relatively large (the order of hundreds of bp).

DOP-PCR: a genome amplification procedure that employs a "degenerate oligonucleotide primer".

Double minutes: extrachromosomal DNA circles containing 1 to 2 million bp that replicate autonomously.

Fluorochrome (also fluorophore): a chemical compound with fluorescent activity (emitting light) upon light excitation (e.g., fluorescein, rhodamine, coumarin, cyanin).

Genetic linkage maps: maps of abstract chromosomal loci based on allele-recombination frequencies.

Haematopoietic tissue: myeloid or lymphoid tissues in which haematopoiesis occurs.

Haptens: small molecules that stimulate an immune response when conjugated to a larger molecule, carrier molecule, here biotin, digoxigenin, dinitrophenol.

Haematopoiesis: formation of blood cellular components.

Hemocytometer: microscope slide designed to enable cell counting.

***In situ* hybridization:** technique used to reveal the location of specific nucleic acids sequences on chromosomes or in tissues.

LD_{50}: median lethal dose; dosage that will kill 50% of the tested group.

Lectin: carbohydrate-binding protein. Some lectins are used as toxins to induce immune reaction and increase the percentage of dividing cells in hematopoietic tissues.

L-glutamine: amino-acid used to boost cell growth in culture media. It is rapidly damaged and must be regularly added in culture media. It is light-sensitive and must be stored frozen.

Linkage group: a group of genes known to be linked, assigned to a single chromosome.

Lipophilic: having an affinity for, tending to combine with, or capable of dissolving in lipids.

Microdissection Needles: capillaries used to microdissect. This glass tubes are pulled from both extremes while heated in the middle part. Gradually the middle becomes thinner and finally cuts. The results are two capillaries with a needle tip in one end.

Mitostatic: substance blocking the mitotic process.

Mitotic index: proportion of dividing cells in a cell population or in a tissue.

Molecular cytogenetics: biological discipline that applies molecular methods to cytogenetics.

Nick Translation (NT): a technique of tagging DNA with haptens or fluorochromes through the incorporation of labelled dNTPs in which DNA Polymerase I and DNase I are used.

Osmoregulation: active regulation of the osmotic pressure in the blood.

Otolith: hard calcium carbonate structures located behind the brain of ray-finned fish, contribution to balance and orientation.

Paracentric inversion: a chromosomal inversion not involving the centromere

Phosphate buffered saline (PBS): buffer solution, isotonic and non-toxic to vertebrate cells, basically composed of NaCl, KCl, Na_2HPO_4, and KH_2PO_4. It can be easily purchased from many suppliers as tablets or as concentrated stock.

Pipette puller: equipment used to make microdissection needles.

Polymerase Chain Reaction: basic technique in molecular biology used to exponentially amplify DNA sequences, obtaining copies of original fragments.

Primary cell lines: primary culture refers to the stage of the culture after the cells are isolated from the tissue and are proliferated until they reach confluence on the available substrate. At this stage cells must be subcultured and become primary cell lines.

RAMP-PCR: PCR reaction where the temperature from one step to another increases (or decreases) gradually according the user setting.

Repetitive DNA: DNA sequences that are present in many copies per genome/ chromosome set.

Sex chromosomes: chromosomes that play a role in sex determination.

Stringency: in FISH, stringency controls the specificity of the hybridization between the probe and its target.

Sturgeon: common name used for fish of the family Acipenseridae (around 25 species).

Synaptonemal complex: proteinic structure that forms between homologous chromosomes (two pairs of sister chromatids) during meiosis and that mediates chromosome pairing, synapsis, and recombination (crossing-over).

Tissue explant: living tissue removed from the natural site of growth and placed in a medium for culture.

Synteny: the condition of two or more genes being located on the same chromosome region whether or not there is demonstrable linkage between them.

Tagged nucleotide: 2′-desoxynucleotide 5′-triphosphate with a detectable molecule attached.

Transposable elements: DNA sequences characterized by the capability to move from one site to another within the genome, frequently duplicated during this process.

Trypsin: serin protease used to help to separate cells from cultured fibroblast carpets.

Fluorescence *in situ* Hybridization: methodology used in molecular cytogenetics to detect specific DNA sequences (target sequence) in structurally conserved cells/chromosomes. It is based on the pairing (hybridization) between the target sequence and a suitable probe (tagged DNA fragment homologous to the target). The product of the hybridization reaction is detected *in situ* by fluorescence.

Zoo-FISH: FISH technique, in which metaphase chromosomes of one species are probed with DNA from another species.

Mainly Used Stock and Working Solutions

Carnoy's fixative
Ice-cold 3:1 methanol (or absolute ethanol): acetic acid, prepared just before use and maintained in crushed ice

EDTA – 0.5 M
18.61 g of EDTA
100 mL of distilled water
Adjust pH with NaOH pellets. EDTA will completely dissolve when pH is adjusted

HEPES/NaCl buffer
595 mg HEPES + 4.375 g NaCl/500 ml D_2O
pH 7, adjust with 1 N NaOH (21.25 g NaOH/500 ml D_2O)

Giemsa (5%) (from solution)
5 ml Giemsa solution, 95 ml phosphate buffer, pH 6.8

Giemsa (from powder)
0.5 g Giemsa stain powder
33 mL glycerin
33 mL methyl ethanol
Dissolve and store in dark bottle

Hypotonic solution (Chondrichthyans)
0.3% KCl, 0.25% sodium citrate, 0.28% NaCl in distilled water

Hypotonic solution (most ray-fin fishes)
KCl 0.075 M
0.558 g KCl/l of distilled water. Store maximum one month at +4°C

McIlvaine's buffer
Solution A 0.2 M anhydrous $Na_2HPO_4.12H_2O$ (60.87 g/850 ml D_2O)
Solution B 0.1 M anhydrous citric acid (3.84 g/200 ml D_2O)
pH 7 823.5 ml solution A + 176.5 ml solution B

McIlvaine's buffer with $MgCl_2$
0.203 g $MgCl_2$/100 ml McIlvaine's buffer
if pH > 7 adjust with solution B, if pH < 7 adjust with solution A

PBS buffer, 20x pH 7.4
160 g NaCl
4 g KCl
36 g Na_2HPO_4
4.8 g KH_2PO_4
1 L ddH2O
Adjust to pH 7.4 and store at RT
It is to be diluted and aliquoted as 1x PBS

PBS, 10x (protocol 1)
0,15 M NaCl
0,05 M NaH_2PO_4

PBS, 10x (protocol 2)
75.8 g NaCl (1.36 M)
9.93 g Na_2HPO_4 (70 mM) or
12.45 g $Na_2HPO_4.2H_2O$
4.14 g NaH_2PO_4 (30 mM)
H_2O q.s.p. 1000 mL and sterilize in autoclave

Phosphate buffer, pH 6.8
Solution A: 8.16 g KH_2PO_4/l of distilled water
Solution B: 8.52 g Na_2HPO_4/l of distilled water
Working solution: 51 ml solution A and 49 solution B, final pH: 6.8

SSC, 20x buffer, pH 7.0
175.3 g NaCl
88.2 g Sodium Citrate
1 L ddH$_2$O
Adjust to pH 7 and store at RT. It can also be sterilized in autoclave
This stock solution is used for working solutions after dilution to 0.1x, 1x, 2x, 4x SSC

Index

L

M

N

P

R

S

T

V

Y

Color Plate Section

Chapter 5

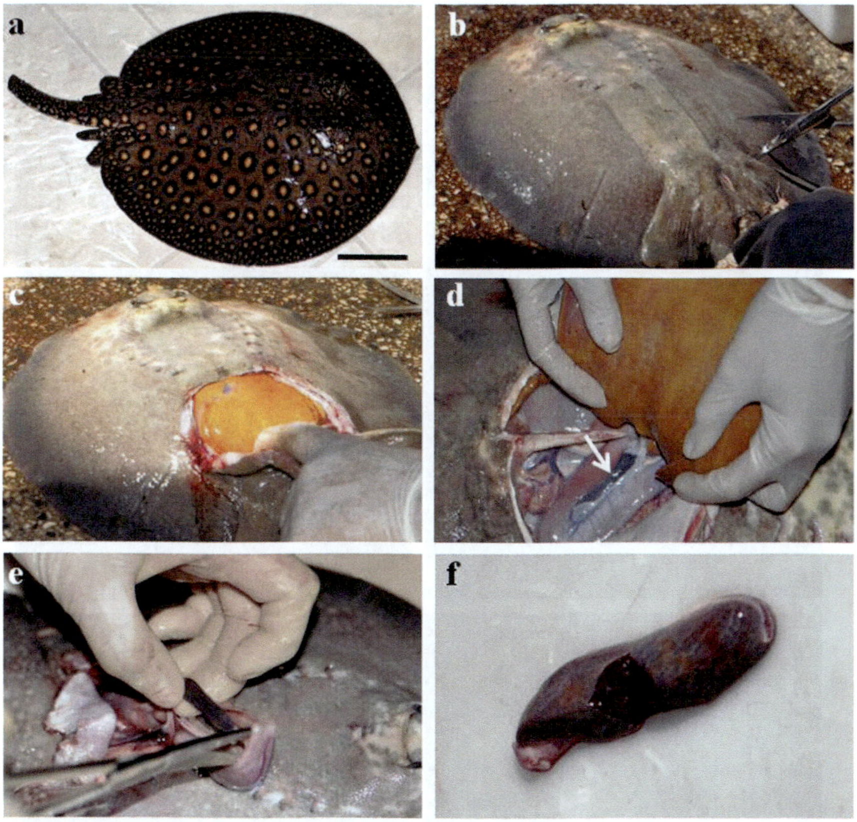

Figure 1. Stingray tissue sampling for chromosome preparations. (a) *Potamotrygon motoro* (male); (b) ventral view of the body; (c) abdominal cavity showing the liver (yellow); (d) removing the liver, organs are displayed; the arrow indicates the spleen, adhered to the intestine; (e) removal of the spleen, and (f) cut the central region of the spleen.

Chapter 11

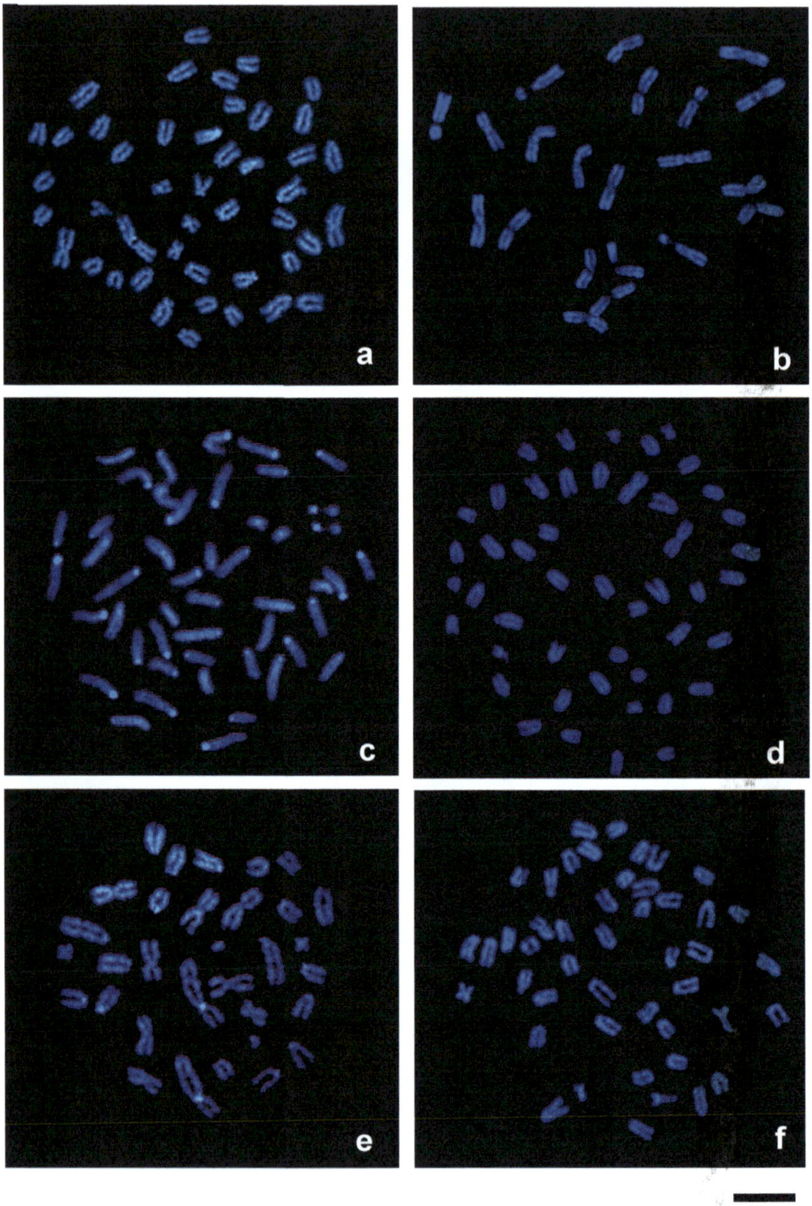

Figure 1. Metaphase plates from kidney and spleen cell culture in Antarctic notothenioid fish species: (a) *Trematomus hansoni*; (b) *Notothenia coriiceps*; (c) *Gymnodraco acuticeps*; (d) *Trematomus newnesi*; (e) *Trematomus pennellii*; (f) *Chionodraco hamatus* (DAPI staining). Scale bar: 10 μm.

Chapter 12

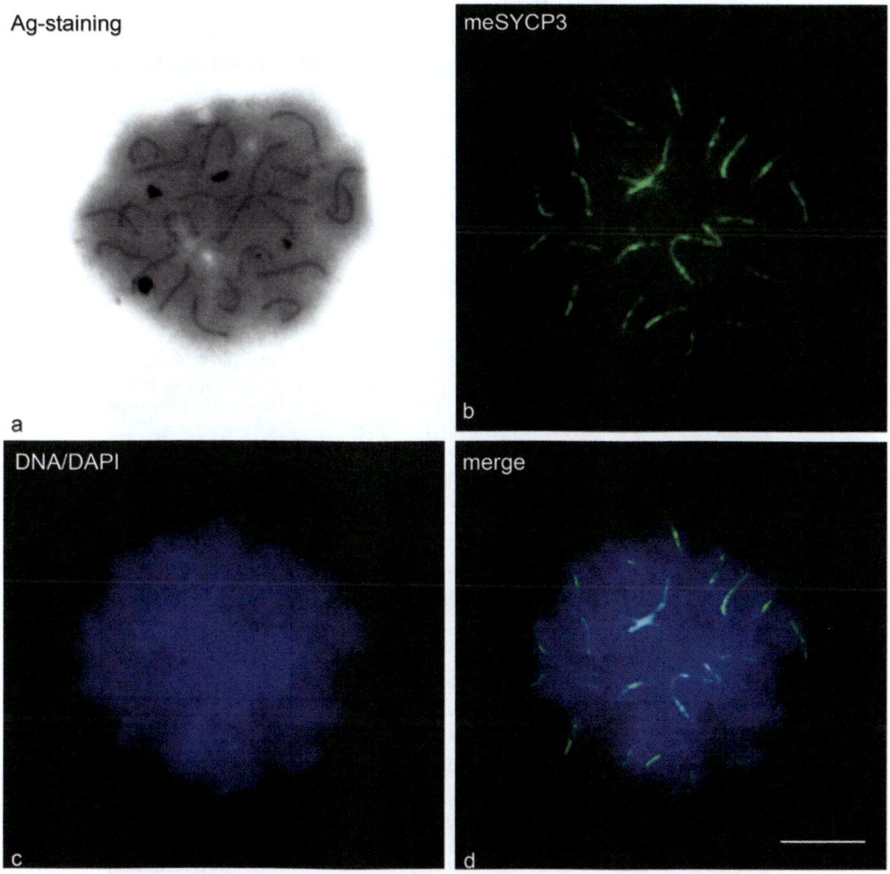

Figure 1. Photomicrograph of an *Oreochromis niloticus* pachytene spermatocyte stained by silver nitrate (a). The stained SC elements are covered by the chromatin of bivalents. In (b), pachytene spermatocytes stained with the anti-medaka SYCP3 antibody. SCs are selectively labeled. The same cell was stained with the DNA-specific fluorochrome DAPI, showing the chromatin only (c). In (d) association of the two images. Scale bar 10 μm.

Chapter 15

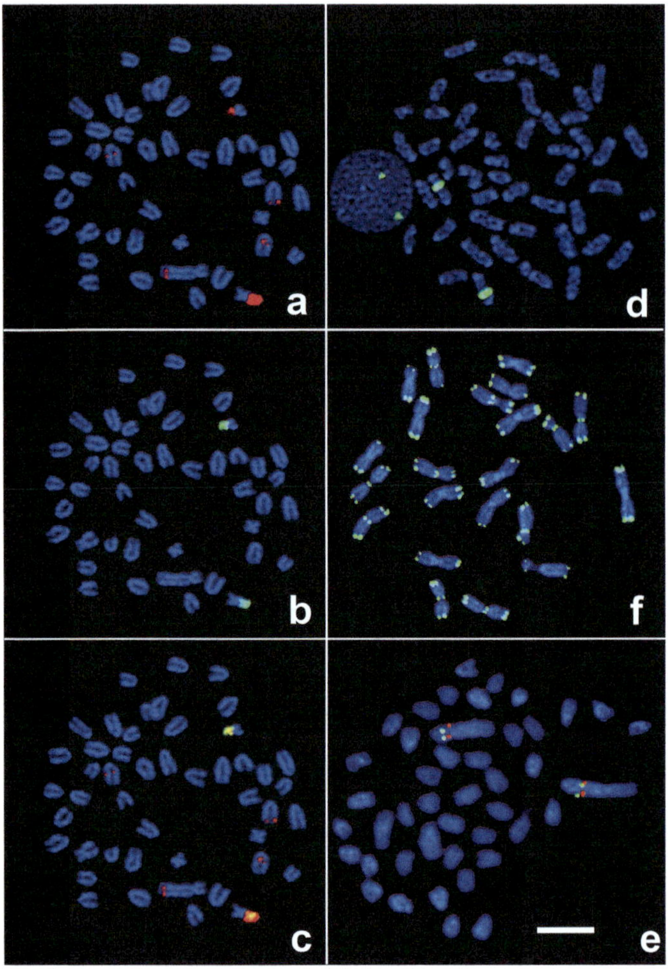

Figure 1. Examples of FISH obtained using probes indirectly labelled with biotin, detected with avidin-fluorescein (green signals) or digoxigenin, detected with anti-Dig-rhodamin (red signals). Double color FISH of 5S- and 28S rDNAs cloned in plasmids on a metaphase plate of *Chionodraco hamatus* (Channichthyinae): 5S rDNA hybridization pattern (a), 28S rDNA hybridization pattern (b), merged image (c). FISH of an immunoglobulin gene cloned in a plasmid on a metaphase plate of *Trematomus bernacchii* (Trematominae) (d). FISH of the telomeric repeat probe $(TTAGGG)_7$ on a metaphase of *Notothenia coriiceps* (Nototheniidae) (e). Double FISH of two BAC clones from the LG3 linkage group on a metaphase plate of *Oreochromis niloticus* (Cichlidae) (f). Scale bar for all pictures: 10 µm.

Chapter 16

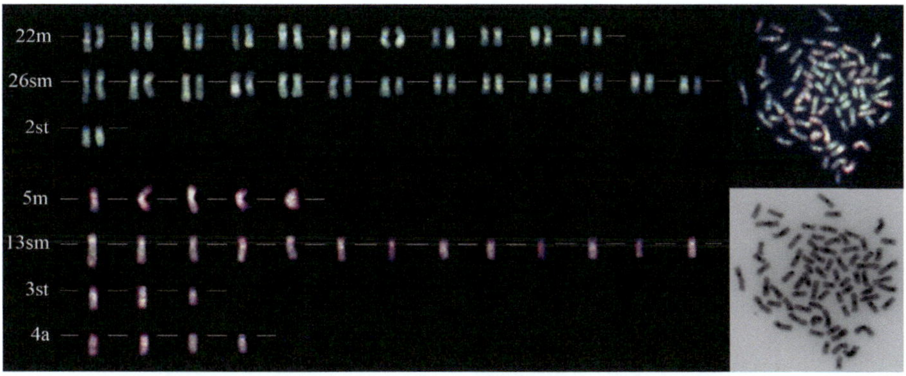

Figure 1. Representative metaphases of the *Cobitis 2elongatoides-1tanaitica* hybrid visualized by GISH with *C. elongatoides* and *C. tanaitica* genomic probes. Homologous and homoeologous chromosomes of triploid hybrids cannot be distinguished in normal DAPI stained metaphase, however, the GISH clearly distinguishes the *C. elongatoides* (shown as those with green fluorescence) and *C. tanaitica* chromosomes (shown with red fluorescence) in *C. 2elongatoides-1tanaitica* hybrid.

Chapter 18

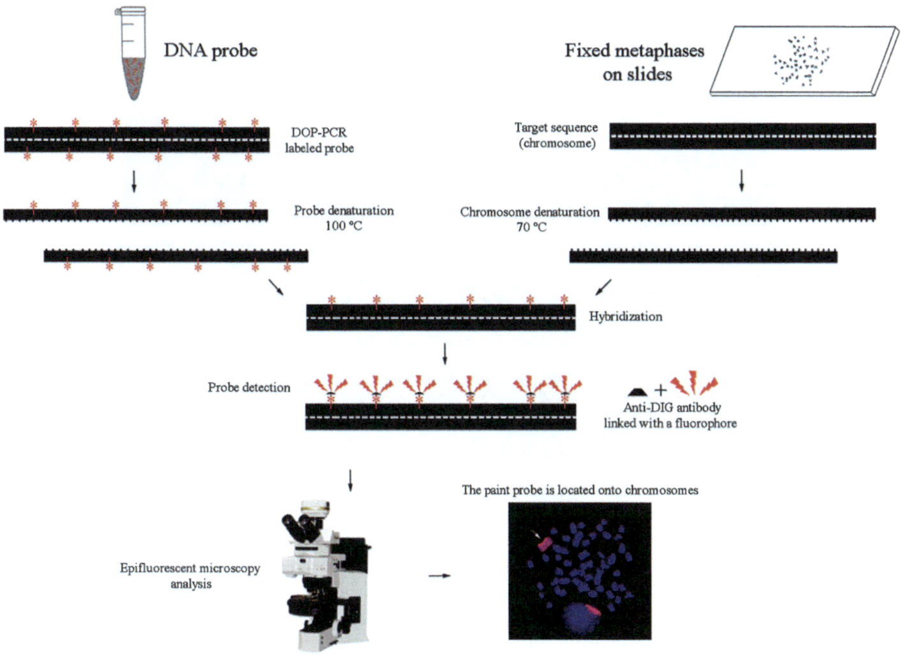

Figure 4. Scheme of chromosome painting with probe obtained from microdissected chromosomes.

Chapter 19

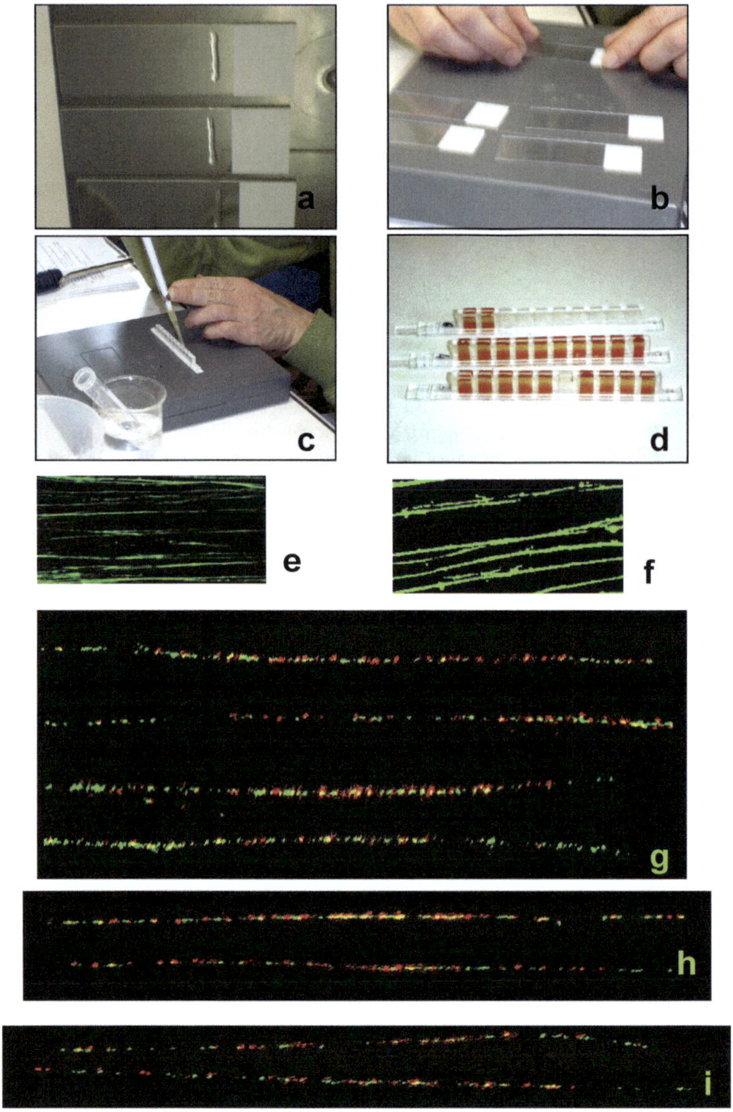

Figure 1. Lines of suspension, along with the frosted part of poly-L-lysine slides (a); dragging the suspension with a large coverslip (b); filling up plug moulds (c and d); YOYO-stained stretched fibers as seen using a fluorescence microscope magnification X40 and X100 (e and f, respectively); double-FISH of a DIG-labeled 5S RNA ribosomal probe (red) and biotin-labeled 28S RNA ribosomal probe (green) on stretched fibers of Antarctic notothenioid fish, as described in § 4, showing alternate copy patterns: *Trematomus bernacchii* (g); *Chionodraco hamatus* (h) and *Trematomus pennellii* (i). For the hemoglobineless channichthyid fish *Chionodraco hamatus*, which has no erythrocytes, we counted lymphocytes, prepared DNA fibers from a concentrated lymphocyte suspension, and obtained similar results.

For Product Safety Concerns and Information please contact our EU representative GPSR@taylorandfrancis.com Taylor & Francis Verlag GmbH, Kaufingerstraße 24, 80331 München, Germany

Batch number: 10726532

Printed by Printforce, the Netherlands